Anthropic Reality

Science, History, and Religion come together to define a never ending reality

By Steve Preston

© Copyright 2014, Steve Preston
All rights reserved.
No part of this book may be reproduced, stored in a retrieval system, or transmitted by any means, electronic, mechanical, photocopying, recording, or otherwise, without written permission from the author.

Table of Contents

TABLE OF CONTENTS 3
INTRODUCTION 6
THEORY OF ABOMINATION 18
CONSERVATION OF EVERYTHING 23
CONSERVATION OF TIME 25
ADDING LIFE 27
LIFTING A CAR 29
WHAT IS LIFE? 31
EGYPTIANS AND FREUD 33
DEMONS 35
JEWISH VERSION 37
THE LIGHT 39
MORE LIGHT 47
VIBRATING ME 51
REALITY 52
8 DEADLY SINS 55
DEFINED LIFE 56
FAITH AND REALITY 59
SPACE RESONANCE 61
LIFE RESONANCE 64
RESONANCE & QUANTUM MECHANICS 66
VIBRATIONS OF EXISTENCE 69
BOOK OF SECRETS 74
CONSERVATION OF EVERYTHING 81
CONSCIOUS COLLECTIVE 83
VIBRATIONAL DISTINCTION 85

WHAT IS A LIFE IN-WAVE?	87
LIFE'S OUT-WAVES	89
IN AND OUT WAVE DIFFERENCES	92
GOING TO ANOTHER UNIVERSE	94
WHERE IS THIS UNIVERSE?	95
SYMMETRY NOT CONSERVATION	97
MICHAEL NEWTON'S SOUL JOURNEY	101
BRAIN CLASSIFICATION	105
LEVELS OF CONSCIOUSNESS	109
NEAR DEATH CONSCIOUSNESS	115
OUT OF BODY CONSCIOUSNESS	117
DEATH IN RELIGION	121
JEWISH DEATH	128
NEW TESTAMENT DEATH	135
AFTER DEATH DESCRIPTION	142
HEAVEN WAR LOSERS	147
DEMONS	151
UP UNTIL NOW	153
DEMON MAKING LILITH	154
JESUS REMOVES DEMONS	155
JEWS TRY TO DRIVE OUT DEMONS	160
RAISING THE DEAD	161
SLEEPING DEAD	164
DO WE KNOW PEOPLE AFTER DEATH?	168
DEAD SOULS?	171
LOSE LIFE TO GAIN IT	172
"GOOD" DISASTERS SHOW REINCARNATION	176
RE-ENTRY OF LIFE	180

MANY DEATHS	184
REINCARNATION VERSUS RESURRECTION	186
DANIEL AND DEATH	189
DEATH FROM NO HEARTBEAT	191
HELLISH HELL	192
HOW DO PEOPLE DIE?	195
AFTER A FINAL DEATH	197
CONCLUSIONS	199
ABOUT THE AUTHOR	204

Introduction

Two things are noted in the title; Anthropics, the science of subconscious control, and Reality; the fleeting condition we believe is around us. In this book we will look at how our reality, including what we believe to be the reality of death can be viewed in a more "scientific way". Once you see what reality is, you can enjoy it more fully, Life can be understood more completely and death does not have to be feared. OK! There may still be a tiny bit of fear, but it should reduce anxiety of anticipation that sometimes fills our alone times. Please don't be scared by the necessary oddness of this book, use of Biblical texts, uncomfortable topics, and the strangeness of 10 dimensions controlling the makeup of our universe. All will be reconciled in the book to a reasonably comfortable level so you can begin to understand death and enjoy life better. As part of the details, the book goes through some of the more esoteric conditions of life such as reincarnation, astral projection and similar things that may help you fill in some blanks concerning what you can't see but you need to understand.

It's hard for us to understand life, because we are living it. It can be harder to understand Death when we are dying it. What I will show you is there is, generally, no real death. We'll prove it with Science, Historical discovery, and Religious detail.

A nice example is of a fish that cannot perceive water because water is a "Requirement" for his reality. We can't recognize our capability to mold reality, because we are living IN it. Just like matter and energy that cannot be destroyed or created. Life is the same. I know you have seen the body of people shrivel up and turn grey and they bury the husk of a person, but we are going to go beyond this simple characteristic and truly investigate what life is and how very important people are to this perceived universe.

From my other works and the writings of many scientists, I think you may know that life is REQUIRED for this universe to exist. The idea that life can, somehow be lost would begin to destroy our universe. Life makes up 3 of the dimensions in what I have been calling the Ethereal Dimensional Dynamo of our universe. While you may not even understand, yet that there are many dimensional "qualities" that are required to make a universe and the old length, width, and height you learned in school as the dimensions of the universe are absurd. In the age of quantum mechanics and Einstein's Theory of Relativity that has shown that distance is "temporary", we know that those old definitions cannot describe the universe at all. We know now that what we perceive as matter and energy are both built around their own special dimensional dynamos but life is special. The simple answer is "We cannot have matter and energy without life", so don't even try to say there are only 3 dimensions. Besides that, we cannot get into a discussion about life and death without two things- Religion and Anthropics. I'm not talking about hard core, fire and brimstone, religion, but it is a backdrop that may help us understand Anthropic Reality.

Anthropics

The anthropic view of the universe has shown that **nothing exists without life being there** to "appreciate it". I'm not talking about Schrödinger's Cat; where they put a cat in a box and added a type of poison and built a theory around the fact that the cat was neither alive nor dead until someone actually opened the box. Nor am I saying a tree doesn't make a sound when it falls unless someone is there to witness it. -------Science is saying it and no one has been able to come up with a more plausible view that is not completely full of issues. You see; anthropic science has found that many of the anomalies we face when looking at the universe can only be solved by stating the occurrence is based on the exact time we are in, with the exact level of advancement, etc. One simple observation is that the stars are generally moving away from the center of the universe where the Big Bang was.

The problem is that the center appears to be on Earth so the earth should not be here. The only way to address both is to have "our version of now" as the beginning and either goes backwards in time from us or forwards from us with us being the center of it all. This is not simply an egocentric vision full of self importance. It seems to be the only way to look at our perception of reality.

Our Soul Is Outside Reality

I know I'm losing you so let me start over. The Theory of Anthropics is not exactly centered on people. It seems to be centered on a component of people we loosely call our soul. A soul is tied to out perception of reality, but it is not exactly IN this reality. Instead, it seems to be joined together with other souls in some way and it is this union that "defines" what we perceive as reality. Let's say this union of souls decided one day to not want the color blue. Not only would there be no blue anymore. There would NEVER have been blue from the beginning of time. Pretty weird isn't it. Fortunately or unfortunately, the more we look into existence the more this oddball notion fixes the problems in definition and understanding. As we go through this description, you will see this special energy called "the light" sometimes. Whatever it is called, simply remember what Einstein said when asked if a tree fell in the woods and no one heard- "did it make a sound?" His answer was "there was no tree". He was not trying to be coy! We can control our reality rather than simple observing it. The more we separate from this reality, the more we control it. The various writers in the New Testament writers told us Einstein was telling the truth. Here are their own words.

Matthew 10:39- *He who finds his life will lose it, and he who loses his life for My sake will find it. Accept what you are able to do and what you are not able to do.*

Matthew 16:25- *For whosoever will save his life shall lose it: and whosoever will lose his life for my sake shall find it.*

Mark 8:35- *For whosoever will save his life shall lose it; but whosoever shall lose his life for my sake and the gospel's, the same shall save it.*

Luke 9:24- *For whosoever will save his life shall lose it: but whosoever will lose his life for my sake, the same shall save it. For what is a man advantaged, if he gain the whole world, and lose himself.*

Luke 17:33 *-Whosoever shall seek to save his life shall lose it; and whosoever shall lose his life shall preserve it.*

John 12:25- *He that loveth his life shall lose it; and he that hateth his life in this world shall keep it unto life eternal.*

You may have read these things without even recognizing the significance to Einstein and Anthropic science. I'm sure many just said this is just to weird and skipped over the message that was so important during the time of God Incarnate that it was stated at least 5 times to insure we would READ IT. What they say over and over again is that if one loves carnal life [that portion we usually see touch and feel], he will not be able to realize what life is really about [and be able to mold "reality" as needed to understand and enjoy it. Another way to describe the message is that dying is not death but release to adapt reality. Let me give you an example everyone has heard about while we are still alive.

Car Lifting

Our soul is really a life energy with "understanding of reality" rather than our "self" which only "experiences reality until the time we call death". This gives the soul "special powers". Take, for instance a girl who sees her dad get trapped beneath a car. While he might die, somehow, even a girl can lift a car off a man and her backbone doesn't snap. It is impossible no matter how many people tell you she just used adrenaline to change reality. What she used was the Biblical "ignoring of carnal life to modify it". I'm sorry I had to burst the adrenaline bubble, but "normal" physics would not let the bones hold for the rescue. We must use anthropic physics. Without it the bones would shatter. Yes, people are telling you the muscles somehow become like rock to protect them, but that is way too weird.

Faith and the Soul

If you have read the Bible, you probably remember when Jesus told his Disciples that if they only had the faith of a grain of mustard-seed they could move mountains or turn water into wine or bring people back to life or walk on water. None of those things can be done without a soul. None of those things can be done in this reality no matter how much adrenaline you have. Jesus showed his followers a way and many of those things were done. Elisha Jesus, Peter all brought people back to life, a bunch of people walked on water, Sticks were turned into snakes and water became wine and people were cured from horrible illness. Here is the weird thing. Many of the people who did some of these things, thousands of years ago and still today, are not religious. This faith stuff was something far different that acceptance of the living God. We'll get into this as we go.

Defining Death

This being immersed in life is one of the main reasons we cannot adequately define life. Like a fish accepting water as being there, we believe life is just there. It's part of our reality. What we are going to attempt in this book is to understand the water around us. The water [or life] is made up of 3 things the self, the soul, and the spirit. This subject is even more difficult to understand, because REALITY doesn't exist without life. Therefore, these 3 components of life are actually dimensions of existence in the truest sense. To define death we must define life as death is simply an extension of life. If you don't believe it right now, hopefully, you will begin to understand as we go along.

Dead DNA

I know all this sounds like craziness, but I am going to get you through the hard parts so you can understand life and with it, you will begin to understand death. Speaking of death, what is the difference between live DNA and Dead DNA? If you give, I will tell you the answer---- "No difference in the carnal existence we perceive". Dead and living DNA have the same sugars, same, interfacing bonds, same structure. The difference is that a dead one is not connected to the union of the souls; so to speak. Without the soul, there is no life. There can be matter and energy, but there is no life.

The Self

The outside of your existence is something we can call the self or the body. It is the part that is experiencing the "feelings" of existence. It is what we call carnal. One definition would be the part of us that is focused on self, survival and sex. I know some of you are saying three cheers for self, but those tactile experiences greatly reduce someone's understanding, and interaction with the truer world. The self must be tempered with the SOUL.

The Spirit

I have not discussed the third dimension of life--- the spirit. Let's get through the first part and then we will tackle that one. It is the simplest and the most difficult one. To be honest, it has more to do with death than life, so I think it is appropriate to hold off a while until we get to death.---- No skipping to the back of the book to get an understanding of death right now. First, we have to establish life by creating what we call reality.

Science and Religion

While science and religion should never be separated as one will totally lose perspective very quickly, most discussions that cross between the two, generally, get bogged down. What I will attempt to do here is go through mostly scientific observation and discussion and follow it up with more religious determination will scientific reasoning. I can't guarantee it won't be confusing, but I think you will gain insight if you read the book.

That being said, discussion of scientific life needs to begin with description of reality. Matter and energy do not need reality for existence, but without reality, both would be different. Let me give you some examples.

Light Force

When we think of photons or electromagnetic energy that vibrates at just the right frequency for our world to come alive with what we call light changed completely when it vibrates faster and become the dangerous XRAYS and gamma rays that shoot though seemingly solid objects with ease. If the same electromagnetic energy slows in frequency and becomes stationary it becomes something we call electricity. I know you think you know what electricity is, but the definition is the "Potential" to become electromagnetic [it's a want to be]. If this same wave vibrates too fast it, become magnetism. I know you think you know what that is as well, but nobody does. At both ends of this activity, everything is black---there is no light and really, there is no perception of the energy, but it is still there.

A Marble

While you are chewing on that, let's look at a marble made of gold. While we think of it as a mass, it actually is a vibrating "substance" called Aether [according to Einstein]. This Aether stuff is kind of like a potential to be a mass. As the mass gets larger, its gravity gets more observable until it gets so large, the mass is no longer observable and it is like total gravity. A mass can be defined as an aethereo-gravitational wave. As it vibrates faster, it becomes different types of mass. At 60 Exahertz, it is perceived as gold. Slower vibrations make less dense things we call mass like carbon, or hydrogen. Slower vibrations make the sub atomic and even sub-particulates like Quarks, bosons, gravitons and similar "sort of non-existing mass". If it vibrates too quickly, it becomes something we call pure gravity. Steven Hawking would call it a black hole. I know you haven't heard about this way of describing mass, but those crazy string theorists and Quantum Mechanics Scientists look at things more mathematical so they can define away. The main thing I'm trying to show here is that what we believe to be reality is not and what we believe to be life is so much less than what life really is.

Collective Red

Reality is made up as a collective. The color red is simply a certain vibration of electromagnetic waves, but every time you and everyone else "sees" this particular vibration, we not only build a color, we create visibility. I'm not saying everyone must understand RED the same way. I'm sure some perceptions of that color are different than mine, but the definition is similar enough for us to all live in this same reality.

Just like perceived matter, photons making "red" start out as something we call electricity. As this electricity vibrates, it begins to have what we call magnetism [just like matter has gravity]. The faster the "electro-magnetic wave" vibrates the more powerful it becomes. Soon it become visible, then beyond visible. As gamma rays is become terribly dangerous and faster than that it can become total magnetism when no more electricity can be perceived.

10 Dimensional Reality

As I have been discussing, matter, without the forces needed to hold it together, is defined by its own dimensions or dimensional dynamo. I don't mean length, height, width, but the actual dimensions are mutually perpendicular in vibration similar to the ones we used to think meant something. Length, height, and width only mean something when viewed from a single aspect with time and distance both as a constant which we now know is preposterous. These, still taught, fake dimensions have limited understanding of existence and life for a long, long time. The way to look at matter, as I just mentioned requires dimensions of Aether, Gravity and a union of the two to build a combining wave. Only when all are held together can something "exist" as a particle. Matter can't exist without energy to hold it together, but it is controlled by its own set of dimensions. It exists as a three dimensional dynamo consisting of electric potential, Magnetic field, and the electro-magnetic waves generated by their combination.

Time

Here is where the 10th dimension comes in as, whatever time is, none of the vibrations of these dimensions mean anything to this world without time. Certainly, time can be defined by more dimensions. The idea that "life" doesn't age as quickly if one if going near the speed of light certainly shows there are more dimensions in time and the idea that God sees existence laterally [sees the beginning and end of time simultaneously] possibly does mean time can be expanded into additional dimensions. That being said; for the purposes of this discussion of life, it might be better to simply think of it linearly. With time being considered a single "glue" dimension, we can perceive 3 mutually perpendicular dynamos---each with 3 mutually perpendicular dimensions of their own as shown below. This book is about the Ethereal or life dynamo shown at the top. Please understand why I chose to have time considered as a single dimension as the inner connecting elements of various dimensional dynamos becomes difficult to draw as the unions and dependencies are addressed. Trust me; having time as a constant [single dimension] will not detract from our understanding of this awesome thing, we call life not will it harm our study of death.

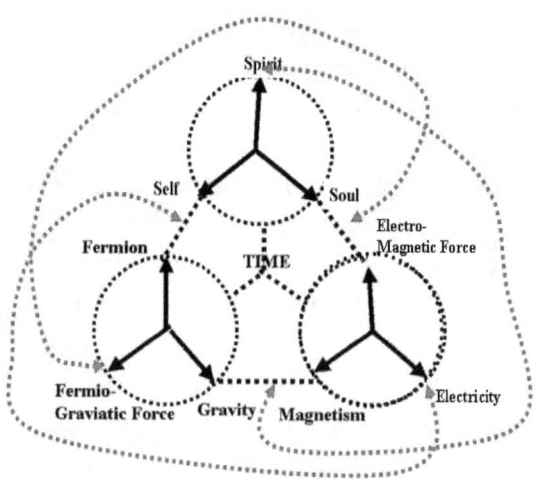

Life/Death Dimensions

That brings us back to the three dimensions of life that turns a lump of DNA that has been energized by the electromagnetic forces of our perceived reality into life. It requires the combination of self, soul, and spirit as described in the Bible.

> *1 Thessalonians 5:23- "And the very God of peace sanctify you wholly; and I pray God your whole **spirit** and **soul** and **body** be preserved blameless unto the coming of our Lord Jesus Christ."*

You probably noticed that Mass and Energy dynamos had something in common. There was a "potential" for it when there was no vibration and when it vibrated too fast, it sort of went "beyond our reality" as pure gravity or pure magnetism. The spirit is that part of life that goes beyond our reality. If you have heard about heaven, this is sort of a tunneling element for that place. The "Self" is the potential for life and the "soul" is what can goes beyond our reality as we venture away for this thing we can call self. Let me stop here and look back at one of those difficult statements presented earlier, but this time with the more acceptable translation.

> *Whoever wants to save his SOUL will lose his carnal life, but whoever loses his carnal focus will find true life"* **(Mathew 16:25).**

Remember that little girl that picked up the car off her dad? This is sort of saying the same thing. Our soul is, *generally*, tied to this reality. This reality is kind of like a governor on a motor that keeps it from going too fast and destroying itself. It builds a reality around something that has been called the "resonance of life". I'll get in this "resonance stuff" a little later but, essentially, everything in the universe is comfortable at a certain Main vibrational level. Remember, if matter vibrates too fast, it ceases to be matter and the same it true of energy. Well--Life is the same. If we [or self-soul wave] vibrate too fast, we will cease to be. That is, we will cease to be in this universe. The faster we vibrate, the more we control our surrounding "reality" just like mass gets more powerful as it vibrates faster and electromagnetic waves become more violent forces like gamma waves, etc. . Everything in our universe is tightly affixed to everything else, but that does not mean everything is RIGIDLY defined.

If we can somehow vibrate faster than the "life reality around us, we have more control over the reality, but at the same time less interest in this reality so it is a two edged sword. On a small scale, you would recognize that if someone is "meditating" he loses his awareness of the things around him. That's what Jesus was stating in the book of Matthew. The person who, for an instant, gains tremendous power to help someone out from under a car is done much the same way.

Peter Walked on Water

Don't believe Peter didn't walk on water because someone said it's impossible. Don't think that you can't leave your body simply because it seems odd. Don't think that when Jesus told his people that faith as small as a grain of mustard-seed is all that is required to move mountains was a lie. He was God incarnate and we can move mountains just by telling them to move if he says we can. In the Anthropic version of physics, it is not only possible, it can be considered a requirement that is stifled by people simply "enjoying" carnal life with its sex, satisfaction, pride, power, and all the things that we begin to lose as we gain power with our soul.

"Light"

In the New Testament, we can read that God incarnate sent down his "Holy Spirit" to allow us to "See the Light". According to the Biblical account, once the Holy Spirit interacted with a person's spirit, he could face death without fear. He could raise people from the dead, he could heal the sick, and he could generally speak for God. The examples of the exploits of those who received the Holy Spirit are numerous. God's assurance in the

Bible is that the Holy Spirit can help ALL who want him to come. The problem is that the 3 S's keep getting in the way. One must abandon**, self, sex, and survival** to allow this Holy Spirit entrance to modify the vibrational level of life, so to speak, and allow a better control over one's reality including the reality of what happens after what we call death.

Only Meditate With Direction

Before we talk about death, let's briefly discuss something ominous. The verse indicates that "only those who lose there souls <u>FOR GOD</u> will find it". By definition if nothing else, life must be indelibly tied to the Creator God [Whatever you want to call him]. The nature, cause, effect, care, understanding, and temper of God can help us understand a lot about our life's happiness and hurt. The ominous part is that if you just meditate on Valhalla without seeking guidance, your energized soul may not find its way to this heaven place unless this Holy Spirit helps guide a person's spirit.

Theory of Abomination

With that let me tell you the very sad story of millions and millions of people called the ANAK. These people tried to rebel against God under the leadership of a certain "Satan". While they originally were beings called angels, this act did 2 things. It first turned them back into humans, but then came the horrible punishment. They were doomed to walk the shadows of this universe after death and could never again be raised up or go to heaven. For centuries, they have been wandering and waiting. Drifting around in some kind of cosmic soup, not dead, not alive, not focused on some direction and existence can become what could be called HELL. If an opportunity arose, these miserable souls would sort of coexist within a person. I have heard of people who have "left their bodies only to have "someone" step into their life. Sometimes the people are able to return. If not, we need to talk about pigs.

Pigs For Comfort

As we switch to pigs, let me tell you about another interesting verse in the Bible. First a little background is in order. Satan and a group of rebels fought a battle a long time ago and lost. He and his followers were turned into humans that were missing something the Bible and other ancient texts called the "LIGHT". This wasn't electro-magnetic vibrations but something akin to the 'spirit'. Without this "light" all Satan's followers were doomed to stay on earth after they died as demons. It was scary being a demon so when Jesus drew some of these demons out of a person; they begged him to let them go in some pigs. Even going in a pig was better than being in the nothingness that the demons and, presumably others who lose their spirit and must wander aimlessly in the afterlife. Some would simply call it hell. I don't know how difficult it is to get inside a pig's life, so these helpless "dead" probably must stay in this state of hell forever.

Before I go on I want to establish a theory of mine I call the Theory of Abomination. As you read through ancient Jewish historical references and Biblical testimony, there is a strange thing that is noted about animals. Very few animals are considered "clean" while many are considered Abominations. The list of abominations is odd. They include porpoise, whale, Ape, pig and many animals we consider to be almost like us today. The main reason they are abominations is that they were made [genetically mutated" by the ANAK and the "Book of Secrets", which we will get to later called the act of creating these animals as horrible or worse because the creators could not understand what creations really required. That is the background and here is my theory. There is a reason the demons we straight for the pigs when they were forced to leave a "human". That reason was that the millions and millions and millions [1/3 of the entire population of the universe called Heaven according to Jewish and other religious texts]. Just imagine all of these souls in misery until they could inhabit "some" type of sentient being. Not as good as human, I believe many of them have inhabited the abominable animals. These are not necessarily souls of bad people, but they did not have something called the "light" that made them cursed and desperate. Don't worry too much about the demons and this weird "LIGHT" right now I will be going over those important elements of life and death as we go along and we will get back to demons later.

That brings us to death. In this book, we will venture into the mystery of death, both good and bad. Surprisingly, there is a vast amount of knowledge concerning death that is both helpful to us while we are alive and, potentially, allows us to almost revel in how death can help us. Before we go on, let me sort of review these general concepts with respect to vibrational frequencies. Possibly, it will make it more understandable.

Particle Variations

This first chart shows how the dimensional elements of matter react with vibrational changes. It should be noted that matter cannot exist without force and life dimensions, but the dimensions making up the Structural dimensional Dynamo control the character of matter as described. Notice the slowest SUB-Particle is Aether. Essentially, it is the potential to become matter. The fastest vibrations cause black holes. Also, notice that gold vibrates with a resonance of 60 exahertz. If you could vibrate items with a touch, you could be like Midas.

Chart of Particle Vibrations

Name or characteristic	Maximum Wavelength [meters]	Highest Frequency [Hertz]
Aether, black matter	$*1 \times 10^{+10}$	$<30 \times 10^{-3}$
Human hearing	1×10^{4}	20×10^{3}
Fermion [part mass]	$*1 \times 10^{+4}$	30×10^{3}
Boson [smallest mass]	$*1 \times 10^{-0}$	30×10^{7}
Baryon [electron]	$*1 \times 10^{-3}$	30×10^{10}
Hydrogen/1	1×10^{-9}	30×10^{16}
Berylium/9	1×10^{-10}	30×10^{17}
Silicon/28	3.5×10^{-11}	8.5×10^{18}
Zirconium/91	1×10^{-11}	30×10^{18}
Gold/197	5×10^{-12}	60×10^{18}
Meitnerium/270	3.7×10^{-12}	27×10^{19}
Straight Gravity [Black Hole]	smaller	higher

Force Variations

This chart shows how the dimensional elements of force react with vibrational changes. It has electricity [the potential to become energy] as the lowest vibrational level. Increasing the vibrations, are pretty well known by all of us today, but they are on the chart just the same.

Chart of Electro-Magnetic Vibrations

Name or characteristic	Maximum Wavelength [meters]	Highest Frequency [Hertz]
Electricity [potential energy]	5×10^{10}	$<30 \times 10^{-3}$
Brain function**	5×10^{7}	6 to 10
VHF [radio]	1×10^{0}	30×10^{7}
UHF [radio]	1×10^{-1}	30×10^{8}
SHF [radio]	1×10^{-2}	30×10^{9}
EHF [radio]	1×10^{-3}	30×10^{10}
Microwaves	2.5×10^{-4}	12×10^{12}
Infrared [light]	1×10^{-6}	30×10^{13}
Visible light	4×10^{-7}	75×10^{13}
X-rays	1×10^{-8}	30×10^{15}
Gamma Rays	1×10^{-9}	30×10^{16}
Magnetism [pure kinetic energy]	lower	higher

** it is highly likely that these frequencies are simply catalyst for much higher frequencies actually used by our brains to store thoughts and images.

Life Variations

This chart shows how the dimensional elements of Life react with vibrational changes. Vibration levels slower than 10^{17} are almost completely CARNAL. As one increases the vibration, he slowly becomes self actualized and begins to have something called selfless love [to give up you life for another]. These things allow people to have some control over their reality. At higher vibrations, one could communicate with God, leave your body, walk on water, and even move a mountain. Notice that experiencing sex drives your vibrational level to something like 300 terahertz, so if you are feeling a little buzz, you know where it's coming from.

Chart of Ethereal Vibrations

Name or characteristic	Maximum Wavelength [meters]	Highest Frequency [Hertz]
General Molecular interaction	*1 x10^{+10}	<30 x 10^{-3}
Unaware Life	*1 x10^{+4}	30 x 10^3
Life Awareness	*1 x10^{-0}	30 x 10^7
Survival	*1 x10^{-3}	30 x 10^{10}
Sex	1 x10^{-9}	30 x 10^{16}
Need for Companionship	1 x10^{-10}	30 x 10^{17}
Need to help others [self actualization]	3.5 x10^{-11}	8.5 x 10^{18}
Selfless Love	5 x10^{-12}	60 x 10^{18}
Universal Understanding	3.7 x10^{-12}	27 x 10^{19}
Insight into external Universe	smaller	higher

*My list is somewhat different that those called chakra levels by the Buddhists, but there is a similarity that cannot be ignored.

No matter how you sense it, increasing your vibrational resonance frequency, increases your power over your environment.

At the very high frequencies, there is almost no need for the environment. Eliminating the lower frequency elements allows you to sustain the higher levels longer and better. What we will find later is that reducing the consideration for self not only allows someone to life a happier life, but also it makes what we call death better.

Conservation Of Everything

Let me get away from religion. We have no issue stating that "Mass cannot be created or destroyed". We also know "energy cannot be created". It can change state, but it is always here. Guess what!!! "Life cannot be created or destroyed". It can change states, but it cannot be destroyed. This is especially true in Anthropic science where the Life dimensions actually control reality much more that the vibrations of Magnetism, Electricity, Gravity and the like which only become real when an entity witnesses it. If you have every heard the phrase that photons are sometime a particle sometimes a wave. Let me tell you something about waves----they don't really exist in what you actually understand as reality, so sometimes photons exist and sometimes they DON'T.
Now I have to bring up quantum Mechanics, even though I hate to. I just said photons sometimes don't exist. The quantum mechanics scientists have proven now that when a sub-particle goes into this "don't exist" state here, they miraculously appear somewhere else in the universe. They cannot truly disappear. While I'm saying all of this, please understand that Einstein and all the other vibrational scientists will tell you that nothing is really here. Vibrating dimensions interacting together make waves that can be useful for these soul things to build a reality.

—

Please take a breath and quit getting upset with the book. Think of flowers or streams or mountains and close you eyes. While they aren't really solid, you simply don't care so you can relax. I only brought this stuff up to describe one thing.

If life cannot be destroyed, the simple answer is that there is no death. Death is an extension of life. There are many parts to death and we will look at them, but never consider death as the end of existence. In fact, there is a cyclic nature to the life/death tango.

Conservation of Time

Time cannot go just one direction. If time was always "advancing" soon all time would leave the universe and there would be NONE. I know that sounds silly, but there is a simple method used in our universe to all for conservation of time, just like everything else. That secret is an adjoining time reversed universe. While some call this universe heaven, it makes no difference to me what you call it. The three dimensional particle dynamo [this is where mass and gravity are generated] causes vibrational flow through the universe according to Einstein and many others. This set of waves is called out waves in that the soon reach the end of the universe and can be lost forever if it wasn't for another dynamo of dimensions we can call the force dynamo [this is where electro-magnetics is generated]. As the vibrations from these [in-waves] coming into our universe to build stresses are generally exactly like the out-waves, except they are backwards in time. Guess what as our particle [out-waves] leave out universe they become "stress in "Heaven" and produce force and the Particles from outside our universe become those in-waves we need so desperately to hold masses together. Let's not look at life for a minute and just try to observe things happening in our adjacent universe. Everything would be going backwards. Things would be getting younger and newer to us as time over there is backwards.

As time leaves our universe, it is sensed as backward time in our neighbor and vice-versa. We always have time and they always have time. String scientists call all this stuff "super-symmetry". I'm not just making stuff up---you know! The diagram below, hopefully, will describe what a more up to date description that these string theorists have put together and seem to work. The universe on the left might be ours while the one on the right might be the one we call heaven. While there are billion of mass vibration centers, I only drew one at the boundary of our universe. As it travels into our neighbor, the same thing [according to super-symmetry] is coming back at us, but everything is backwards so it forms stresses in our universe that we call Electro-magnetic forces that hold everything together and allow particles to build.

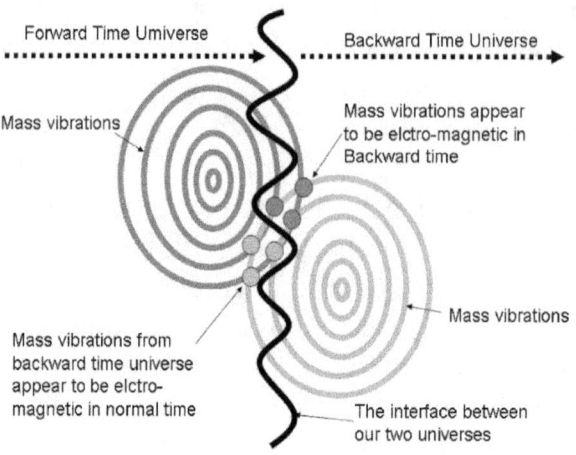

Each one of the dots drawn where the two vibration rings intersect can be considered a particle seed. From these, additional vibration rings will emerge which represent the combination of the out-waves and the in-waves.

All mass is made this way.

By the way, things would not feel backwards to people living in the adjacent universe if you were worried about them. To them, our life starts at its end and has us getting younger and younger.

I know you were confused when Einstein told us that if you go near the speed of light, you quit aging and some of you understood that going faster than that would mean you would go backwards. This is the same concept, but scientists differentiate backwards time as an undefined reaction as density goes to infinity along the reduced speed timeline. One way to look at it would be that particles become electromagnetic ways above the speed of light.

Adding Life

Please don't shut the book. I know I keep pleading, but you will learn something in this book. I Promise!!! Now for the fun part; if "Life" leaves this universe, it will be in the adjacent one and a "new life" will be generated in our universe. If this were not so, NO NEW BABIES COULD EVER BE BORN. Conservation of life would suggest—life cannot be created and yet babies are born everyday.

The bad people are saying," See there must be death or life cannot be added." You mean spirited people have just jumped to conclusion. The Bible states that when you die, the "self" turns to dust and the "spirit" goes to heaven. Then it makes a horrible claim that the soul may live forever or not. With all these spirits leaving the universe, certainly spirits are returning as they leave the adjacent universe. These spirits allow for life to be "RENEWED" if a "live" soul is hanging around, the soul attaches, the self sort of is the interconnection of the DNA to make it alive in "this reality" and viola! The sperm and ovum union established a bridge to reincarnate the soul into what we call CARNAL life.

When I said there is no death, I meant the soul continues to live and controls the universe. The soul part of a person is not solidly linked to this CARNAL world, but can go beyond this world. The simple example was that car lifting trick I mentioned before. Here are some examples of this actually happening. The soul was able extend itself and in the words of Jesus, move a mountain with the "Faith" of a grain of mustard seed.

Soul Collective

Many believe there is a collection of souls that are waiting for reintroduction as a new person following death. The Bible is a major source of confidence this is how life energy exchanges occur. When someone's "Self" dies, the life element of him or her can do a number of things. Here are a few.

Sleep -- Whenever Samuel was "AWAKENED" by the witch of Endor in the Biblical reference, he was very upset that he was being disturbed.

Inhabit other living beings that already have a "self"- the demons of the Bible could introduce their souls into people and into pigs. Many other examples show this to be a very real concern.

Can be awakened and readied to be introduced into a new "Self"- This is called reincarnation and it is described over and over again in the Biblical texts.

Can be reenergized such that dead people could be brought back in their own bodies. Certainly, this has a time limit as the body decomposes quickly when not continuously renewed from some drive given by a host soul. Many of the ancient Jewish people recorded this as well as others. Elijah, Elisha, Jesus, Peter and many others were able to bring people back to life. Others around the world have done this feat as well and there is something important to know about these events. The reenergized people do not remember where they were. This shows us the initial attribute of "perceived death" is sleep. We will see the soul is awakened later, but let's give it some time.

We'll get into these and also show the collective soul control over our universal reality and how we can modify this reality and do seemingly impossible things. We'll also begin a discussion on what an introduced soul "Remembers" to allow it to be considered an entity. This is a somewhat touchy subject, but we still need to discuss it.

Lifting a Car

I'm thinking you are skeptical about the lifting a car thing so, before we go on, I thought I would give you a short list of reports of this "impossible" type of interaction in recent times. This is only a small number of the real quantity, but it will show you how common this type of event is.

Adam Simmons was working on his daughter's Jeep Liberty when it fell on him. His 22 year old daughter, Rachael, lifted the 5,600-pound car up to free her father. He escaped with only minor cuts and bruises due to his daughter's quick action. She had no injuries and her backbone stayed in place.
In 1982 Tony Cavallo of Lawrenceville, Georgia, was repairing his Chevrolet Impala from underneath. You guessed it. The vehicle was propped up with jacks, but it fell. Cavallo's mother, Mrs. Angela Cavallo, lifted the car high enough and long enough for two neighbors to replace the jacks and pull Tony from beneath the car.
In 2006, Tom Boyle, of Tucson, Arizona, watched as a Chevrolet Camaro hit 18-year-old Kyle Holtrust. The car pinned Holtrust, still alive, underneath. Boyle lifted the Camaro off the teenager, while the driver of the car pulled the teen to safety.
In 2009, Nick Harris, of Ottawa, Kansas, lifted a Mercury sedan to help a 6-year-old girl pinned beneath.
 In 2011, Danous Estenor, of Tampa, Florida, lifted a 1990 Cadillac Seville off of a man who had been caught underneath.
In 2012, Lauren Kornacki, of Glen Allen, Virginia, rescued her father, Alec Kornacki, after the jack used to prop up his BMW slipped, pinning him under it. Lauren lifted the car, and saved his life.

In 2013, in Oregon, teenage sisters, Hanna (age 16) & Haylee (age 14) lifted a tractor to save their dad pinned underneath.

Tom Boyle Jr. saw a crumpled frame of a bike under a Camero bumper, and tangled within it a boy, trapped. The bicyclist was pounding on the car with his free hand, screaming. Boyle bent down, grabbed the bottom of the chassis, and lifted it off the trapped boy.

The reason I'm bringing these up is for 2 things. First, to let you understand that the stories you have heard about this type of thing being done are not flukes and, while impossible, they happened and the people doing these feasts were ordinary and they were not harmed. Don't even give me the explanation of adrenaline. What happened in these cases is that they vibrated themselves to a level that could change "normal" reality. They completely ignored themselves, so there bodies went outside the limitations of our CARNAL world. I know it sounds weird, but think of it as the next step beyond Maslow's "Self Actualization" where cosmic understanding is achieved by caring for others.

What Is Life?

Before we get into Death let me give you sort of an image of life made up of three distinct parts- the Self, Soul, and Spirit.

Self

What people see and your carnal character, sexual interest and attraction, physical self, body, pride, hunger, vanity, smartness, the almost unnatural feeling for self preservation, and the awareness that you alive are all parts of what we can call the SELF. These all make up a veneer that interacts with others in a physical way.

Soul

The soul, on the other hand, is the real you. That statement will take some explaining. Sometimes called the ID or subconscious, the soul reacts with and **can sometimes modify how we perceive, sense and combine realities**. The neat thing about this "part" of you **is it doesn't die** like the "SELF".

Spirit

The "spirit" can be considered as a glow. It is a window between this world and beyond. The part of us survives beyond reality. The diagram following is one I typically use to show you an initial impression of our three entities.

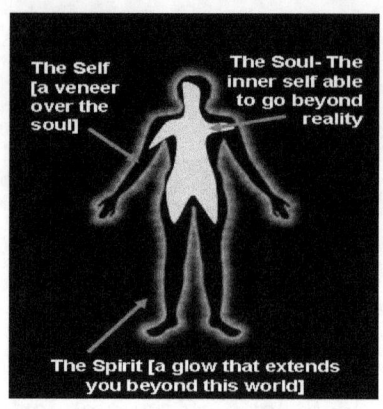

The answer to what our life is won't be simple, nor can it be defined in a simple image.
1. <u>The Bible says dust to dust</u> when referring to the self or body. When our body dies, it is no more. Don't go thinking that is the end, but when you are renewed, you will be different.
2. <u>The Bible says, our soul can live or die after we die</u>. This will be described in more detail as we go along. The warning is this- "What profits a man who gains much but loses his soul?"
3. <u>The Bible indicates that your soul may have several bodies</u> before our spirit is released. Reincarnated souls are described in the Bible and many other ancient texts around the world.
4. <u>The Bible also indicates that the Spirit part of us helps us to interface with God</u> and allows our soul to live in this place called heaven after death.

Like Christianity, most religions identify the soul as the real entity of man. I think Taoist say it best by telling the followers that if they are to be happy, they must think of themselves [the self entity] as an empty vessel. The Bible teaches the same thing it just says it in a slightly different way.

"If you want to live, you must die"

This isn't dying in the way we normally think of it. What the book tries to tell us is that if we try to enhance our Carnal SELF, we will lose our Soul. Losing your soul is a true death, so let's not let that happen.

Egyptians and Freud

The Christian "3 person in one" is universal in the ancient world and pretty much in today's world, so I hope none of this is too strange. The Egyptians had the Baa, Kaa, and shadow while Sigmund Freud had his Ego, ID, and Superego. The problem with all of them is the emphasis on the Self or the Baa or the Ego. The emphasis of these carnal elements is what can doom a soul to misery. I'm not talking about a lake of fire, here, I simply talking about the wonders that can be achieved with the faith of a grain of mustard-seed. Don't ignore the soul. Don't ignore the most important part of your existence. It is the part that won't die.

Conservation of Energy
I think I had better go over this again. According to Einstein, vibrational nothingness associated with mass and light continued outward from a central point and escaped the limits of the universe at infinity. This means that sooner or later, all will be dark and no mass or energy will exist. While that would certainly be in the distant future, it still makes me uncomfortable and it really doesn't make sense that existence is not renewing. Additionally, conservation of Energy laws would be violated. As the light energy hits the end of our universe and leaves, it is IMEDIATELY replenished. Everything else in the universe renews and we can, pretty much, be assured that the same is true of mass and light [electro-magnetic stress]. Everything in this universe is continually renewed.

If energy is renewed, <u>life must be renewed</u> in a similar way so we can gain information about ourselves by looking at non-living energies as well.

Conservation of Time

Time must also be rejuvenative. Here's the deal. Time must go backward while it is going forward. I know that sounds weird. This will be an interesting concept when we look at death in particular. Does death come first and a person gets younger or is it the other way around? Are there just snapshots in time and only the now is real? If the now isn't real, are we as "selfs" real? Maybe Chromosomes can help.

Dead Chromosomes

A dead chromosome and live one look the same and have the same characteristics. The sugars are the same, the links and bonds are the same. Something outside the chromosome makes it alive. This thing outside the DNA and Chromosomes is what we can call life. In particular, it is the "self" driven by the "soul". Both seem to be "Conserved" in this universe. As one set of chromosomes or a person becomes dead, another seems to become alive. I know you are thinking there are more people alive today than during the old days, so this "Conservation of Life" doesn't look like it will hold water. This is because you are not identifying life without self. Just because a soul has no body does not mean there is no life. That brings us to demons.

Demons

By far these demons are one of the most frightening components of life and death and, I'll bet, you don't know what or who demons are beyond the little bit I presented so far. Most don't even worry about such things. Sure, they are in the Bible, but that was thousands of years ago. All of those Exorcist shows are Hollywood things, and mental illness is a disease of the mind.

Demons have been around since the ANAK people began dying thousands of years ago and--------------they are still here because of an ancient curse from GOD.

Lilith Leader of Demons

About now, I typically begin a discussion from Jewish texts concerning Adam's first wife. Her name was Lilith and she became the mother of the demons after handing out the fruit of the tree of knowledge to Eve, Adam's second wife. Here is the same story from an ancient Moslem text that I think you will find just as enlightened and it shows that the Jews were not the only ones afraid of the woman who became the "succubus".

"Our Father Adam"

At the first Adam was male and female in one body, man on one side, woman on the other. In due time the female part separated from the male, and became a perfect woman; and the couple mated. The female refused to submit to the male, saying they were made of the same dust, and he had no right to order her about. So she was turned out of Paradise, and, consorting with Iblìs **[Satan]**, *and became the mother of devils. [She is called "El-Karìneh" by the Arabs, and* **"Lilith"** *by the Jews generally; "El-Brûsha" by the Sephardim, or Spanish Jews].*

*She is the deadly enemy of all women, especially such as have recently become mothers. --When "El-Karìneh" had been driven from Paradise, Allah created our mother Hawa [that is **Eve**], out of one of Adam's ribs, which He had extracted from the latter whilst he slept. Adam and Hawa were very happy together till Iblìs succeeded in getting back into Paradise concealed in the hollow of the serpent's fangs [This serpent was none other than Lilith, or in this story ElKarineh]. Having entered the garden, Satan succeeded in persuading Hawa to eat of the forbidden fruit, which, according to some of the learned, was **wheat.** Adam, having been persuaded by his wife to share his offence, was, as a punishment, cast out of Paradise, together with Hawa, Iblìs, and the serpent. He had, however, the sense to snatch up, and bring down to earth with him, an **anvil, a pair of tongs or pincers, two hammers, and a needle**. ----- Two hundred years elapsed before Adam and Hawa met once more ----- Hawa had borne offspring of the seed of devils [sons of ElKarineh], and Adam had got many children by female jinns [daughters of ElKarineh]. The descendants of those unclean monsters under the name of afrìts, rassad, ghouls, marids, and so on, still people the earth and try to harm mankind.*

Jewish Version

"Our Father Adam" goes on and is insightful, but I need to step back a little to show the similarities. The Jewish version is slightly different, but the results are the same. Lilith, somehow, could turn herself into a serpent. After her mean act, she was cursed to "Never leave the earth again nor her children". Additionally, there was another major group of people who had been cursed the same way before all this serpent stuff.

Her partner in crime Iblis [or Satan] conspired against God and initiated an ancient war. The outcome of that war was the destruction of heaven and earth. It was a nasty war. As punishment for the insurrection, all the rebels became humans and their souls were NEVER allowed to leave this universe.

As Lilith's descendents died, they became, sort of, the walking dead as did the followers of Satan [called the ANAK]. With all the souls walking around, they needed leaders. Satan was a natural leader and Lilith became a leader as well. For the "demons", it has not been a nice time. When Jesus forced the demons out of an unfortunate man, they begged him to send them into a herd of pigs just so they could experience a tiny bit of what we call reality.

Without a self, the prospect of getting a self, and without the curtain of sleep, these demons walk in terror and misery until they can inhabit people. Sometimes it only makes people insane, and other times, a person may not even notice the encroachment. After all, most do not want to disturb the main host or they would get kicked out.

Once they are kicked out, they sort of drift in between time. Think of it this way. Suppose you could not make out images of what we call light, but instead you simply saw the real vibrations. What if you could not hold or feel, or hunger, or anything? I know you are thinking that would be like hell, but that does not have to await any soul.

This is not a religious book so get that right out of your head. Unfortunately, there will be a lot of religious texts presented to help confirm the science.

Just because there is science in understanding that a soul that cannot relate to a perceived reality is in danger of succumbing to misery, does not mean religion, it just means scientific theory based on massive amounts of ancient text, and modern scientific research. We'll get back to demons later.

Vibration to the Rescue

Here's the thing! A soul can go outside what we might call the resonance of the unified soul cluster by simply vibrating faster. The poor soul of Lilith's descendents and the followers of Satan have a problem. They lost something that is called the "Light" as one of their punishments. If they had this light stuff, they could have more control in vibrating away from a carnal sub-existence.

The Light

For this little chapter, I need to direct you to the first verse in the book of John in the New Testament. Certainly, there are many ancient texts that can help here, but I just like this one. Most likely you will see that it is saying something a little different than you had believed before because you skimmed over it or perhaps you don't read in the Bible. What we find is that "light" and the spirit is the same thing. Finding out what the light is may help us understand the whole SPIRIT thing which will help us understand something about death. Here are basic things to see about the "nonphotonic light".

1. This light and spirit seems to be closely connected.
2. Reanimation and reincarnation or life can occur. Apparently, even without the Light, reincarnations occur, allowing for additional Carnal living by a soul for some not well known reasons.
3. With this light, a transition from this world to the one called Heaven is possible.
4. The Light was introduced into Adam and eve and all of their descendents.
5. According to ancient texts, if someone had children outside the Adamic bloodline, they lost this light and became something called gentile.
6. Soon almost all people had mixed bloodlines and were gentile.
7. This is where Jesus, God incarnate reestablished this Light to allow the previous actions of the spirit/ light.

I'm sure most of you disagree with me on some of these points, but the main thing is the Light/spirit was extremely important to our DEATH. Following are a few of the many ancient texts describing aspects of this important part of our being.

"Life of Adam and Eve"

[Gnostic Text Chapter 18.1] Eve said to Adam: *"Long may you live, my lord to you is my life submitted, since you did not take part in either the first or second collusion. But I conspired and was seduced, because I did not keep the commandment of God. Now <u>separate me from the light of this life</u>."* [This Adam and Eve not only had this special LIGHT thing integral to them for a time, but also they knew about it. After the Apple eating, Adam says he should not be allowed to have this Light. The Light was removed from the Rebel "Watchers who turned into the ANAK" after losing the first Heaven War. It was resident in the Adamic people and allowed Noah and his children to be called the "Chosen Ones".]

"Apocalypse Moses"

This is another Gnostic Text Chapter 33:2. *"And she [Eve] gazed steadfastly into heaven and beheld a <u>chariot of light</u> borne <u>by 4 bright eagles</u>- and Watchers going before the chariot. In this section, [Eve watched Adam being taken into heaven on a chariot of light after his death. This light comforted, and rode with him in death].----Chapter 36- And Seth telleth his mother, that they are the sun and moon and themselves fall down and pray on behalf of my father Adam. Eve saith to him: '<u>And where is their light and why have they taken on such a black appearance?</u>'* And Seth answered her, *'The light hath not left them, but <u>they cannot shine before the Light of the Universe</u>, the Father of Light; and on this account <u>their light hath been hidden from them</u>.* [This was talking about the light thing that had been taken from the survivors of the first heaven war. The discussion about these people not being able to shine has to do with their cure. They would roam the earth after death as demons.]

"Enoch"
This book was described in Biblical texts, but was lost to early Bible compliers Chapter 56:3. *And the righteous shall be in the light of the sun. And <u>the elect in the light of eternal life</u>:* [This is talking about getting the "Light" back if the Holy Spirit of God was integrated into a person. With this new light, one has an opportunity to experience a different type of life after death.]

"Adam and Eve II"
This is one of the Essene Texts Chapter 22: 8. *"When Enoch had ended his commandments to them, God transported him from that mountain to the land of life, to the mansions of the righteous and of the chosen, the abode of Paradise of joy, in light that reaches up to heaven; <u>light that is outside the light of this world</u>; for it is the light of God, that fills the whole world, but which no place can contain. Thus, because <u>Enoch was in the light of God, he found himself out of the reach of death</u>; until God would have him die.* [This begins to paint a more complete picture of this light that is somehow associated with Life.]

"Hypostasis of the Archeons"
This is one of the Gnostic Texts. *"And he said, "If any other thing exists before me, let it become visible to me!" And immediately <u>Sophia stretched forth her finger and introduced light into matter</u>; and she pursued it down to the region of chaos. And <u>she returned up to her light</u>; once again darkness [...] matter. ---"And he said, 'Come, let us create a man according to the image of God and according to our likeness, <u>that his image may become a light for us</u>.'* [The last verse is very important as the rebels who had lost during the Heaven War continuously tried to regain this LIGHT stuff so they could attain some higher level of life and its associated death.]

"Popul Vuh"
This is from the Mayan Bible. *"Then there is the matter of human beings, whose sowing in the womb will be followed by their <u>emergence into the light at birth</u>, and whose sowing in the earth*

at death will be followed by dawning when their souls become <u>sparks of light in the darkness.</u>" [Sparks of light in the darkness emerging at birth is saying the sparks reenter life after death.]

"Book of Creation"

This is still another Gnostic Text. *Samael [Satan] said, "I have no need for anyone-it is I who am God, and there is no other one that exists from me"---Pisitis [God] was filled with anger and said, "You are mistaken, Samael, <u>there is an immortal man of light</u> that has been in existence before you, and who will appear amid the creatures you have made, and will trample you, and you will descend to the abyss--- then he and his followers made a great war in the seven heavens.* [The War of the 7 heavens is what I am calling the Heaven War described in our current Bible. The Man of light, or man that still has the light, is describing Adam in this passage.]

"Secrets of Enoch"

This is still another Gnostic book Chapter 31:4- *Because Adam was to be lord on Earth to rule and control it, the devil as a fugitive made Sotona [War] on heaven, thus <u>he became different from the other watchers</u>.* [This shows that the losers were different than normal watchers/angels. The difference was this LIGHT thing. If we call this the spirit; that would be a close description.]

"Book of John the Evangelist"

You guessed it, another Gnostic Book. *My father [God] changed his [Satan's] appearance because of his pride and the <u>"light" was taken from him</u>. His face became like a heated iron wholly like that of a man".* [The old "heated iron man head" thing and another simple description of the truer light associated with DEATH.]

"Isaiah"

Let's look at Chapter 14:15 again. *Yet thou [Satan] shalt be brought down to hell, to the sides of the pit. All they shall speak* *and say <u>unto thee, Art thou also become weak as we? art thou become like unto us?</u>* [This is what losing LIGHT does to you. You become weak as other people. Therefore LIGHT makes you or your SOUL VERY strong.]

"Mishaf Resh"

Bible of the Yezidi tells us the same thing- *Before the creation of heaven and Earth Ali dwelt upon the sea. Then Ali went up to Heaven and solidified it. <u>Out of His essence and His light He made six gods</u>.* [This "gods" word is talking about what we call angels or watchers. Ali/GOD's, LIGHT thing, made living souls.]

I guess you understand what light is now. Don't be sad if you don't the whole concept of light as a component of death is one of the very difficult elements of explaining all this to you. For sure, it is something we need to sustain ourselves in a less carnal world and Satan's followers lost their portion when they tried to take over Heaven. One of our goals in life must be to get this light stuff so that depiction of hell, whatever it is, will not be realized. Typically, we don't do well in getting the light the first time around, so God has devised a method for us to "Learn" if we try. The learning may take many "lifetimes" so reincarnation was born. After many reincarnations comes something called Resurrection. I'm not getting into the Resurrection so much, but I think we can look at both together to gain a little more insight. I selected another short series of verses that, I think, help us understand more about this elusive element of living/ dying, and living again.

John 1:1-4

*In the beginning was the Word, and the Word was with God, and the Word was God. -- In him was life; and the life was **<u>the light of men</u>**.*

It first states that the carnal part of out creator GOD [in here called the "word"] was the thing that made the life… This is really talking about carnal life and as we read farther, it states that this carnal life was able to be 'enjoyed or witnessed because of the LIGHT. Just assume this "light" or spirit is actually the ability to go between this "reality" and a reality associated with our "joined universe called HEAVEN in the Bible [String scientists simply call it the super symmetric universe mate, but what you call it doesn't matter. After death, the spirit or light leaves only to be returned to another human to allow the same thing repeatedly. In the book of Enoch we find that when the rebels of the Heaven War who followed Satan had this "Light" taken away which forced them to wander this universe after death without a way to go to our neighboring universe.

John 1:5
And the **light shineth in darkness; and the darkness comprehended it not**.

Then it states that this "light" stuff shines or become real, but the ones in darkness [people not attuned to this light or having it taken away] can't comprehend or use this KEY to heaven. Again, Satan's followers and the descendents of Lilith who were cursed in the same way could not understand or use the light and wander in endless darkness after death.

John 1:6-8
There was a man sent from God, whose name was John. The same came for a witness, to bear **witness of the Light--**. *He was not that Light, but was sent to bear witness of that Light.*

Then it goes on to say that John [called the Baptist] was trying to tell people about this light and how people could get more out of life if they realized this light was with them. Think of it this way comprehension was to interact. In the anthropic science or vibrational matter science, the "Interaction" was to increase ones vibration to allow understanding. Certainly, this is hard to do as all the souls sort of vibrate at what we call a "resonance" [at a frequency that allows all to interpret "reality the same way."]

John 1:9
*That was the true Light, which **lighteth every man that cometh into the world**.*

First, it says there is a false and true light. Both "bring" people into the understanding of reality as they are born. Someone following direction of those "souls" who ever stricken of the "Light" might very well be the false light director. It is saying don't listen to the voices of those souls [or demons].

John 1:10
*He was in the world, and the **world was made by him**, and the world knew him not.*

This reality is held together by this "light" that surrounds us that was MADE by the son of God. Not having "light" from the "Word" [portion of the Creator God] means you cannot comprehend reality as we understand it. This is the reason these demons [those without the light] try to experience life and reality by co-mingling or taking possession of "normal" people. They are completely miserable outside, without ability to "comprehend" this reality.

Demons Again

I am certainly not getting into verse 12 here because that is too controversial, but you get the picture. This chapter is telling you about the misery of demons. These souls are "sort of alive" and "sort of" dead. They only live by infection. This does not mean you are going to wake up tomorrow and be insane, but I would not be surprised if you think someone guided you the wrong way some time in your life.

Don't get me wrong. Demons are not, necessarily, what we would consider bad, but they are desperate and that can sometimes cause a bad attitude. Verse 12 of John 1 indicates that even these unfortunates can "regain" the light, but that is another story and I want to get back to science.

Science simply says life is a vibration or nothingness just like matter and electro-magnetic forces. Anthropic science goes a little farther in that it suggests that a collective grouping of the "souls" actually mold reality to a common desire. If one wants to change, simply ignore the norm. The Bible said it this way. "With the faith of a grain of mustard seed one can move a mountain". I know it sounds simple so if you completely understand life and death there is no reason to continue reading. If you still have questions, there is a lot more data on demons and the Light. That help shape our reality. Here are just a very few that discuss a different type of light.

More Light

If you are convinced that this LIGHT stuff is not an important part of Life/Death, let's read on.

Genesis

Genesis 1:2-4-"The earth became without form, and void; and darkness was on the face of the deep. And the Spirit of God was hovering over the face of the waters. Then God said, "Let there be light"; and there was light. And God saw the light, that it was good; and God divided the light from the darkness.

[This was before the sun light was made in verse 14, so let's try to understand just what this segment might really mean knowing it has nothing to do with sunlight.

- Verse one indicated there was a previous world before the great Heaven War.
- Verse 2 indicates the earth became "void" of life from the war. The books of Isaiah and Jeremiah both confirm the horrible results.
- The "Darkness on the face of the waters has nothing to do with oceans and sea, but is talking about "places with life giving water". Just because a planet has water does not mean there is life.
- Then it says God said "Let there be light" and he separated light from darkness. The losers of the war lost this light stuff as can be found throughout the Bible and many other texts, so this is saying he did not allow the losers [called the ANAK] to have this light stuff.

Iranian Text

This comes from the sacred Zadspram- *"From the seed which was the ox's, they would carry off from it and the **brilliance of*** *light was entrusted to the angel of the moon in a place that seed was thoroughly purified by **the light** and was restored in its many qualities."* [This segment was after the watchers had corrupted almost all the animals. In order to reconstruct the animals, God had to put in more "Light".]

"Revelation of Moses"

This comes from the Gnostic book **"Revelation of Moses"** Chapter 33: 2-*"And she [Eve] gazed steadfastly into heaven and beheld a **chariot of light** borne by 4 bright eagles- and watchers going before the chariot.* [In this section, Eve watched Adam being taken into heaven on a chariot of light after his death. One interpretation is that Adam was able to use his "Light" after death to go into the universe we call heaven.]

"Book of Abraham"

This verse comes from the **"Book of Abraham"** chapter 4:3- *"And they said, let there be light, and there was light. The gods **comprehended** the light.* [This is an expansion of the Genesis statements. Remember this is before the sun was remade. The **odd part is the comprehend word.** It seems that it is suggesting light was more than something visible that all could comprehend.]

Prayer for Enlightenment

In the catholic **"Prayer for Enlightenment"** we find *"O Holy Ghost, divine Spirit of light"* [The Holy Ghost was introduced by our Creator to allow understanding and return of the "light" to humans.]

Chosen Ones

This is a somewhat important point when discussing our perceived death. It seems that almost all humans had lost this "light" because of intra-breeding. Only the "Chosen Ones" survivors of the worldwide flood carried the "Light" as a useable feature of their "life". It was so important for the Chosen One/Jews to not marry outside their pure blood lineage that 621 laws of separation were introduced in the books of Deuteronomy and Leviticus. The Jews ignored them and, generally, all people were what some call Gentiles [not of the chosen ones] by the time God Incarnate came to earth to die, be raised, and reintroduce a Heavenly LIGHT so that souls could make the journey to the next universe and live there.

"Origins of the World"

This is another Gnostic work. *"The troublemaker that was below them all destroyed the heaven and his Earth. And <u>the six heavens shook violently</u>; for the forces of chaos knew who it was that had destroyed the heaven that was below them. And when Pistis [one of God trinity] knew about the breakage resulting from the disturbance, she sent forth her breath and bound him and cast him down into Tartaros [Hell] and when they had become disturbed, they made a great war in <u>the seven heavens</u>. Then when Pistis Sophia had seen the war, she dispatched seven archangels to <u>Sabaoth from her light</u>. They snatched him up to the seventh heaven."* [This has the Heaven War, the troublemaker {Satan} and this odd thing called the light.]

First she [Eve] was pregnant with Abel, by the first ruler [Adam]. And it was by the Watchers that she bore the other offspring [Cain]. -the first mother might bear within her every seed, being mixed and being fitted so that the modeled <u>forms might become enclosures of the light</u>," [This seems to indicate that the losers of the Heaven War and Adamic humans had children together so that the mixed breed might have this light thing.] -------

"And he said, 'Come, let us create a man according to the image of God and according to our likeness, that <u>his image may become a light</u> for us.' [The losers of the Heaven War believed that man would somehow get them back the "light" that they lost in the war.] ----*His intelligence was greater than that of those who had made him. And they recognized that <u>he was filled with light</u>* [This was talking about some type of luminous essence in the descendents of Adam.]

"The blessed One, sent, a helper to Adam, luminous Epinoia [holy spirit] who is called Life. And she assists by teaching him

about the way of ascent. ----But <u>the Epinoia [Holy Spirit] of the light</u> which was in him, she is the one who was to awaken his thinking. [The spirit inside the descendents of Adam was somehow associated with LIGHT.]

"Jubilees"

The book of "Jubilees" is still considered canon in some orthodox Bibles of today. This comes from chapter 2:9*]* - *"Nor may we take revenge on him because he has <u>stripped us of the</u> <u>"light"</u>. He marked out the borders of the world and created man in his own image with whom he hopes again to people heaven, with pure souls."* [Not only note that without the light, the losers of the Heaven War could not take vengeance on any of the heavenly host, they lost some substantial power without this light thing. Also, note that the word "again" is put in the verse to let us know that man was here before Satan's Heaven War and it was recreated after it was over.]

I'm sure some of that was unsettling and may make you think I'm trying to interpret these texts the way I want them to be interpreted. This is always true, but the reason I amplified some areas was to begin you thinking about parts of your "YOU" as different so you won't be afraid of death and so that you can lean about it.

Like everything else that exists in this universe, this light" is made up of vibrating nothing. If you want to describe yourself you might say the vibrating me. If you want to define what we believe to be reality, just say the vibrating reality. Stresses on any vibration in the universe causes the vibration to change. Changing the vibration of reality would change reality itself. When Jesus and dozens of others made water turn into wine, this is what was done. When Elijah and Elisha and the disciples raised the dead, this is what was done again. Don't get me wrong concerning faith in God Incarnates life after life gift, this is how this "life" can be changed.

Vibrating Me

When one talks about people vibrating, it is both simple and almost impossible to understand. Everything---I mean everything, is made of vibration. This includes all matter, all electromagnetics, all nuclear energy, all photons, even <u>all life forces and those we would consider dead...</u> That by itself is not enough information. The second thing is that live and "dead people" control what we believe to be reality. Let me make a simple observation. ---The color RED--what is it? The answer, of course is vibration and more specifically, electromagnetic vibration just wiggling all over the place. While it is wiggling at a certain rate, other colors and things that are completely invisible are vibrating at different frequencies in a constant time observation. Why in the world do we see RED when it is only vibrating nothingness??? The answer has to do with that Anthropic Universe things I mentioned before. When I say anthropic, I don't mean the "seems to" definition people sometimes attach which completely destroys the truer meaning.

Anthropic means that our linked consciousnesses define observations to vibrations. [Notice I did not say "seems to". In fact, our group consciousnesses invent ALL of the [carnal] reality. Carnal meaning the reality we see, smell, hear, feel, etc.

This invention of the souls is what we call reality!!!

Reality

If you have ever heard the terms "Power of Positive Thinking", "Think and Grow Rich", and all other concepts of the 70s which tried to convince you that how you consciously view reality will affect reality, are not only true, they affect your death as well. In the Anthropic World if you have faith of a grain of mustard-seed, you can move a mountain, as Jesus said thousands of years ago and you can walk on water as demonstrated by Elijah, Elisha, Peter, and Jesus so many years ago. With the Anthropic Principle, science and religion can act as a single tool for us to understand God, the universe, and ourselves.

The dead are not the only souls building our reality. Live people souls just like dead souls can shape and mold reality, because there is little difference and what we assume to be reality is manufactured. Let me give you an example.

Eating

Today we know that what something looks like changes greatly what something tastes like. If you see mashed stuff and it tastes crispy, the mind quickly determines what it would like it to be and viola' it ACTUALLY becomes "that" to the taste buds, touch sensors, and emotion centers determining that you like of don't like what you are eating and level of satisfaction you feel after feeling the substance. The taste bud only looks for a chemical so it can vibrate differently. One wiggles from salt, another from a sour, etc. nothing really determines taste beyond these minor elements. The wiggling is in the form of chemical combinations allowed by structure. If one of the "particles" can attach to the crystal, it gets bigger and vibrates fasters making an electrical differential which is "felt" back to nerve centers in the brain and it is magically determined if it was a good or bad taste. Have you wondered how in the world some people "love" the taste of nasty stuff and you can't stand it? If there was a reality to taste, all would either like the brain response or not. Taste is simply not real.

The same can be said for seeing colors, sensing Aether, looking at beauty, etc. It is all pretty much fake and only "defined" by how we interacted with this "reality" over the years.

Being

What is the perception of being? One answer is seeing the things around us and interacting with them. A problem is that there are no real things around us but simply vibrating nothingnesses as Einstein and others have proven. The forces established by the vibrations give us the "perception" of mass. One can assume that even after the soul is released, it could conjure images of reality, especially when there are groups of these souls that are together. Later we will examine to souls between lives if they are not in a state of sleep, but let's continue that live reality.

Electricity and Sight

For this discussion, let's look at sight. Vibrating of something we defined as electro-magntivity makes a chemical change in our rods and cones of the eye. The chemical changes produce something we call electricity which excites portions of the brain. Everyone uses this electricity stuff to define everything, but it doesn't exist. It is a "potential to do something" by definition. It does not exist except for something we call work that requires some outside intervention with this "invisible potential". It is only the magnetic field produced as the electricity changes that makes it real to us. While it is changing or "VIBRATING" it is in our perceived world. The brain remembers what it perceives from the changing electrical signal and simply "defines" what we call sight. Just think what our world would be like if we could REALLY see what enters our eyes. The things we think we see are just reflections of some external light source rather than the actual object. When the external light source is removed, the objects we see are removed from view.

Try to remember that the "real" reality comes from a thing you can't see called the SOUL. The entire universe is controlled by the joining of souls all defining away.

Our existence or what we perceive as our existence is a combined implication of all in existence. This includes existence when we are DEAD.

---

8 Deadly Sins

Some identify Life simply as the small portion of life that can be called the"Self". The "self" is the portion of the person that deals with Self, Sex, and Survival. Absolutely, that part of a person can be killed and experience DEATH.

The Bible defined the soul with a weird word- "FAITH". More precisely, it called the <u>use of our soul</u> in the carnal world **faith**. Other religions have their own definitions, but the answer is always the same. If someone is only focused on using his "Self", "EGO", or body to exist, he or she is destined to be isolated and limited in this world.

7 Deadly Sins Help Define Self

While there are really 8, everyone remembers those 7 deadly sins-- Gluttony, Wrath, Sloth, Envy, Lust, Greed, and Pride. If you wanted to have a reference about what the Self or Body was, these 7 items cover a substantial amount of its characterization. There is another part we call survival, but I think you get the picture.

7 Virtues Help Define Soul

The items called the 7 virtues describe something about the soul-- Chastity, Temperance, Charity, Diligence, Patience, Kindness, and Humility. Many other descriptions could also be added. The main thing is that it is <u>looking outward</u> instead of looking towards yourself.   The Self is the inward looking entity and the Soul is the outward looking entity making our existence. This reckoning is and was so obvious, that many of the ancient texts reveal this separation of conscious and subconscious or Ego and ID or whatever we want to call it. Let's get a perspective from the Egyptians.

Defined Life

The Egyptians essentially said the same thing as the Jews except they called the dimensions of the person the Kaa [body], *baa*, *sahu* and *akh* [the soul or independent attribute], and the Shut [Shadow] the part belonging to the nether world. The image below shows how they would draw these components. The Sahu is the flying part completely free of the body. The Kaa is shown in the jar, while in this reality, it is completely shut out from the "Real" existence. The shadow was, sort of, in between.

Egyptian Book of the Dead on the Ka/Body/Self- *The Osiris X, may he rest in peace, knows* **the names of your ka**, the **aspect of your soul** *that abides in the ground: Nourishing ka, ka of food, lordly ka, ka the ever-present helper, ka which is a pair of kas begetting more kas, healthy ka, sparkling ka, victorious ka ,ka the strong, ka that strengthens the sun each day to rise from the world of the dead, ka of shining resurrection, powerful ka, effective ka..*

The Sahu or Soul Egyptian Book of the Dead on the Sahu/Soul-*I go round about heaven and sail in the presence of Ra, I look upon all the beings who have knowledge. Hail, Ra, I who goes round about in the sky, I say, O Osiris in truth, that I* *am the **Sahu of the god**, and I beseech you not to let me be driven away, nor to be cast upon the wall of blazing fire.* [Like the Jewish description, the soul part of a person can be lost and even be sent to hell]

Egyptian Book of the Dead
"The Osiris knows the names of your Sahu, the form in which you travel our world - the sun. Sahu pure of body, health-embodying ba, ba bright and unharmed, ba of magic, ba who causes himself to appear, male ba, ba whose warm energy encourages copulating." [This description is exactly what I have been talking about. The soul lives beyond our body or self.] That brings us to the "Shadow".

The Papyrus of Nu on Ba and the Shadow
O *mighty One, when he is adored, great one among* ba*s, greatly respected* ba *inspiring the gods with awe when he has appeared on his great throne: then may he prepare the path, justified, his ba,* ***and his shadow****, may they be well provided for.* [Unlike the body, the shadow was not bound to the grave and could go where the body could not. In New Kingdom, tombs it was at times depicted leaving the body accompanied by the *ba*-bird.

Egyptian Book of the Dead Chapter 92
*Let not be shut in my soul, let not be fettered my "shadow", let the way be opened for **my soul and for my "shadow"**, may it see the great god,* [Clearly the shadow was not the soul.] These concepts were slightly different that those of the nomadic Jewish people, but I think you can see the similarity. In modern times, people have struggled with the definitions because it seemed to give them less control over their environment rather than more. Sigmund Freud, for instance, tried to redefine the elements of life into his own concept to try to make it seem that this reality could hold the essence of the three dimensions of life.

Freud

Sigmund Freud tried to connect the differences in characterizing a person without using ancient religion to guide him. He came close, but he missed important aspects. In Freud's model of the psyche there were the ID (instinctive unconscious), the Ego (organized, conscious), and the Superego (moralizing, not entirely unconscious) form an interactive framework which work together in the mind.

Here is what he had to say. *"One of the fundamental functions of the Ego is Reality Testing – reaching into the real world to see if what is believed to be the case actually proves out – but this does not bear full fruit until the Ego has become Autonomous... substantially set free from inner conflicts between the ID and Superego."* [This is sort of backwards from all other descriptions. In his description, the EGO or self controlled the ID and Super-Ego that continuously were at war. A couple of examples are shown next.]

The ID was the evil characterization while the Superego was close to the definition we must place on the Spirit portion of the body; sort of the HOLY component of a person. Unfortunately, he got the ID and EGO backwards. If any portion of our self would be considered evil, it would be the Self portion or our conscious mind. Of course the ID or self is not specifically evil, whatever evil is. It is simply carnal. With that as a background, let's get into a new description of our universe so you will see how everything has a place and everything is required for our universe to operate.

Faith and Reality

I know I have laid out this anthropic word and told you generally about how the universe is modified by people and you have totally believed what I have been saying because it makes so much sense. You are skeptical. At best, you looked up a sight on Anthropic Universe and found out that one way of looking at it is to sense that the universe was simply made for us. God grabbed the universe and modified it so that people could be created. All the creationists roared, but that is simply not the end and if you try to define people that way you simply have something like mashed potatoes oozing around to fill limitations of the universe.

What anthropic physics really shows is that people can modify the universe [to an extent].

Over time, the universe, quite naturally is shifted to be in line with the needs of people. I'm going to tell you how this also affects what you call death, but you first need to broaden you awareness so the details will be more useful to you.

Faith is not Faith

Remember, Jesus told his disciples that with faith of a grain of mustard seed one could move mountains. Here is what he did not say. "<u>Faith in me</u> will get me to move mountains." Surely, he could do that whether we had faith or not. If he wasn't talking about faith in him, what was he talking about? That, my friends, hopefully, is becoming more evident as we go along. Buddhist monks, for instant, have substantial amounts of faith, but they have no regard for Jesus, God incarnate. These monks have done miraculous things. The Gurus seem to have something we could identify as faith, but they also have no specific faith in Jesus. Many of the faith healers around the world don't profess any specific religious order. The Egyptian magicians Jannes and Jambres had no faith in God, but they could turn a stick into a snake. On and on we could go. Faith, as discussed by Jesus in his plea to his followers was something besides faith in Jesus.

I am certainly not saying don't have faith in the living God and God Incarnate. That is a different subject. This is simply saying Faith allows us to change what we might call Space Resonance.

Space Resonance

I think I have you worried right now with the first description of how matter and electricity are correlated. Let me back up a little and restate resonance for this application in the words of Dr. Milo Wolff who is one of the leading master physicists who has greatly extended Einstein's initial work into a "usable" platform. My comments are in "bold".

*"Resonance is composed of a spherical IN-wave which converges to the center **[of the universe and comes from a different universe as a component of the operational dimension dynamo]** and an OUT-wave which diverges from the center **[of the universe and makes up what I call the structural dimensional dynamo]**. Their separate amplitudes are **[close to]** infinite at the centers. **[Like all other resonance factors in the universe, how close they are to being infinite can be considered the "quality of resonance".]** When combined, the two waves form a standing-wave which has a finite amplitude at the center. The standing wave **[appears]** to be the structure of the electron. The inward and outward waves **[sort-of]** provide communication with other matter of the universe. Spin of the electron is a result of the reversal of the IN wave at the center to become the OUT wave."*

While there are still limitations, this, this definition helps us interpret how an adjacent universe "establishes" resonance in this world. The more we communicate with an adjacent universe the faster our vibrational resonance becomes and its quality rises.

Quality of Resonance
Let me explain this "quality of resonance" a little because it is this Quality that will allow one to expand how a soul interacts with

reality. In electro-magnetics, quality of resonance describes the difference between the affect of a circuit outside its resonance frequency and that which can be described when it is in resonance. If a crystal is excited with a vibration that is half of the frequency it likes, it may vibrate a little and nothing more, but if it is hit with the vibration it likes, it begins to self oscillate substantially. Just think of a tuning fork and how it always sounds the same when struck. That is its resonance and the longer it makes the sound describes its "quality" of resonance.

In the electro-magnetic world, this "quality of resonance" depends on many things including what the crystal is attached to, how well the crystal is cut and how homogeneous the crystal is. In the electron or particle world, the same things can be surmised. Purity of the particle and the things that surround the particle affect how close to infinity the standing wave appears.

Resonance and Matter

I guess you are wondering why I even brought up this resonance in the first place, but **resonance holds matter together, it holds time together, and it holds life together**. If enough electrons are in an area that are sensing similar in-waves, they align together just like a crystal. One could say that atoms are resonant plugs that are held together by "like vibrations". Scientists found these things called <u>gluons</u> which seem to act in opposition to other particles and quasi-particles. Gluons hold quarks together. Three quarks and an unknown number of gluons are called an electron. **Gluons are quasi-particles [fermions] that have a negative gravity.** That is, the farther the quarks move away from the gluons the STRONGER the gluon attraction becomes. It is sort of, like the quarks are inside and invisible piece of matter that has a gravity. The closer they get to the surface of this invisible piece of matter, the more the gravity affects the quarks. The center of what we call a gluon would be the resonance point of what an electron whose resonance is defined by the vibrational characteristics of its component parts. I know that was a mouthful so let me state it differently.

Gluons are not odd, they are simply invisible. One can say that they are an in-wave and out-wave collision; sort of the core of an atom.

Life Resonance

Like I told you at the beginning, we are going to go through the science stuff first and then get confirmation from religion, but I want you to see that both ARE the same and I'm going to let you in on a secret. Consciousness/ life/ and death all act the same. If we wish to affect the universe more and have a higher quality of resonance, we must become pure and surround ourselves with things that allow this pureness. OK! I can't exactly define what pureness is, but prayer and meditation is probably more important to our quality of resonance and our capability of affecting the universe than one would initially believe.

Who Cares About Resonance?
Why have I even brought up resonance? If a vibration node gets larger or smaller shouldn't matter to us. Right?

Well----we need to care for a number of reasons. Here are a few.
1. **The higher the resonance** of electro-magnetics, the closer it comes to being pure magnetism. Even at the somewhat lower levels of resonance, depicting light is considered the **most useful** form of electro-magnetics to the universe.
2. **The higher the level of resonance** of a particle or quasi-particle, the **more useful** the matter it becomes. The higher vibrations mean larger and larger particle including god, uranium and the powerful nuclear materials. Eventually matter become total gravity with no real nature associated with mass.
3. **The higher a person's resonance is**, the more he or she **can affect the universe** and their own characteristic universe becomes more in sync with everything else. The closer one can get to God. Buddhists indicate this change in resonance of

a person's "life" as chakra with the highest level allowing someone to completely separate from this carnal world.

Meditation and concern for others rather than yourself raise the resonance of your "Life" which puts you more in tune with the "spiritual" reality. That is a good thing.

Death is still going to be hard to define, so we will have to look at quantum mechanics a little.

Resonance & Quantum Mechanics

I know you are wondering if I will ever give you a definition and understanding of Death in this book, but I'm trying. You cannot simply blurt things out or they may not seem true. In order to allow for you acceptance, I need to present you with intermediate definition and quantum mechanics will provide a step towards definition. It won't be 100 %, because scientists, generally, have no idea what quantum mechanics is. Let me give you sort of a bird's eye view. It will allow us to understand life and death better.

Niels Bohr [1885-1962] tried his best to mess up our minds. He indicated that there is random vibration associated with everything, but if you test the spin of an object, and it is going one direction, there must be another that is indelibly linked and going the opposite direction. If that wasn't odd enough, he indicated that the linked objects could be millions of miles apart when the transposition occurs. Einstein indicated that the whole quantum mechanics theory was junk science that violated all of his theories. Einstein died before tests were devised that generally proved the phenomenon. --- Not only had the linking be seen, but also this elimination of time space as a boundary has been witnessed.

Today, scientists have witnessed sub-particle teleportation that eliminated time-space.

Without "space", vibration has a difficult time being defined. Its definition needs the introduction of a space based REALITY. Before he died, Niels was writing something entitled "Light and

Life Revisited" what he had finished was published after his death, he understood how very close light and life truly are and this quantum mechanics thing may hold a key to help us understand it as well. My version of light and life is a little different than Niels, and it has a lot less math, but the idea is similar.

Eliminate Time and Space

I know you are saying--- "What in the world has he been saying?", but I think some of it will begin to make sense as you go though this book and miraculously, you will understand Death better. You may also be saying, "What in the world causes the elimination of time and space in quantum mechanics?", so here is one theory. In-waves that introduce force and out-waves that establish physical characteristics are each associated with "times" that are backward to each other. While this allows work to ensue, it does something very important besides that. **It eliminates time during the exchange.**

Simply put-- backward and forward time together make the absence of time and quantum mechanics can be more easily understood as these two "times" come together. Putting it another way; quantum mechanics requires a second universe to neutralize time.

Because there is no time generated during the exchange, the exchange cannot be recorded by us. Even if the exchange took a thousand years, there would be no recognizable time for the action. Here is the scary part. Even if one were to go the speed of light there would be no recognizable time.

What?????!!

[Sorry for that last question mark and the 2 exclamation points, I was getting frustrated.] Here is the leap of faith. If there is NO TIME, there is no space. The actions would be resident at any or all locations at once. This, I believe, is where Einstein began nodding off. He was old and tired and unwilling to even consider the elimination of space. Neils Bohr was still young and thought about all of this weird stuff.

What this does is--- allow us to transfer- to anywhere- anything----at least the essence of anything--- instantly just like the Star-trek hyper drive.

The reason one could travel the distance is that there would be no distance so long as one could somehow introduce in-and out-waves together. That is where resonance comes in.

If someone wants to introduce a modification of space-time, all that is required is to temporarily change ones resonance.

Resonance in the Old Days

Jesus, his apostles, Moses, Elijah, and others from the Bible would have used this simple technique to turn water into wine or blood. Simply change one's resonance and something has to give to correct the shift. Do it right and things seem to be miraculous. Don't get me wrong about what I'm saying. As I stated before, Jesus **was and is God incarnate**, but while he was on Earth he was 100% human and he taught some to understand what he meant when he said "With faith of a grain of mustard seed **anyone** could move mountains."

He could just as easily have said that changing your resonance by elevating your awareness to the non-carnal universal of heaven would allow one to control many carnal things simply by removing time-space during the transition for this equalization known as quantum.

I guess his apostles looked at him weird enough when the moving mountains comment was made, so he tried to keep it simple.

Vibrations of Existence

We know that the faster the photon of light thing vibrates the more powerful it becomes. Soon the fast vibrating photon thing becomes dangerous to humans as it can go right through the body [x-ray] and if it slows down too much it changes into something we call radio waves. I know you are thinking that these radio waves must not exist because they don't produce light and they have no mass, but let me assure you that sometimes these photon things do act like normal matter. If you look at the diagram following, there is a wiggly line. The faster wiggling represents a prime particle vibrating faster and faster. Radio waves turn into light that turns into the deadly gamma rays.

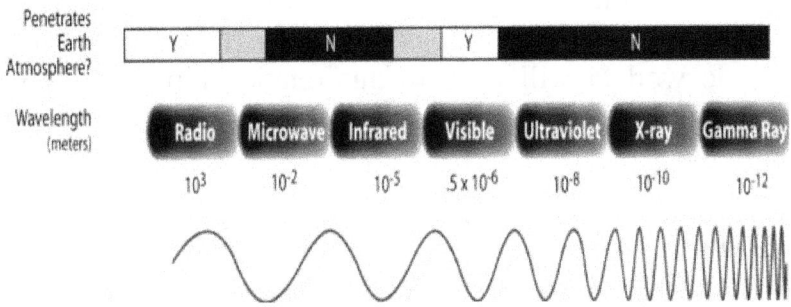

To keep the various limits of these dimension things from seeming crazy, scientists came up with 2 words [Static and Kinetic]. Static components of dimensional dynamos are those components, Like Electricity that have the potential to affect our universe while pure kinetic components of dimensional dynamos like Gravity and Magnetism can cause massive disruption without combination of some of the Static components.

Light

Today, the chart can continue even farther in both directions as light starts its vibrational journey as <u>static electricity</u>. As vibrations start to increase, we call the outcome electro-magnetism. When we get to the highest vibration, we can define the vibration or thing as pure <u>Kinetic Magnetism</u>. All three dynamos have similar static and kinetic limits and dualities with the other 2 dimensional dynamos.

Life and Matter are Similar

As the ethereal dynamo goes through the same type of transformations from the **static carnal life** to the **kinetic soul life** and matter goes through this same transformation from the static Aether up through the pure kinetic gravity of what we call a black hole. We must get a sense that the 10-dimensional universe concept seems to pull everything together.

Resonance of light and Life

It is reasonable to assume that ultrahigh frequency [Kinetic] light, black holes and the soul are joined by vibrational similarity as are the complete carnal life of a tree; the Aether that could eventually make matter and the potential electric fields [Static] that could eventually do work. I will get into this relationship more as we investigate life/Death in more detail, but here is where we are right now.

For light to provide a resonance and be sustained, the surrounding vibrational characteristics of the other dimensions must be similar and this is how our soul can modify reality.

Elimination of Gloom

Let's go slowly. One way this might be attributed is that when we sense the light as **"warm or comforting"**. This is because mass, Force and [your life] are in tune. If you bring in gloom, the same sights will become "Gloomy". In fact, reality around you will be more "depressed". The "power of positive thinking" isn't magic, it is resonance. As you vibrate your soul to a higher, [less selfish, less, carnal way] you will be happier and reality around you WILL be brighter. When a majority of the life-forms are resonating at a similar vibration level, they are unified with a perceived reality and each can interact with the other. Becoming too debased and selfish, or too holy changes everything. Those on the debased side are pulled along barely changing reality in any way. Thomas Maslow looked at the other end and found that people who come out of the gloom can become "Self-actualized". By not thinking of self, the self becomes more in control of the environment.

If we can vibrate faster, awareness of a higher level of light can bring comfort, insight and understanding. This sounds like all that "power of positive thinking" stuff and it does have some similarity, except that vibrational base is what we call life instead of light. Life and light are very similar if you have not noticed.
What I mean by this is that visible light is a moving thing. It goes beyond sight, and extends into comfort, desire, understanding, and awareness. Light even can explain matter because all dimensional dynamos are intricately locked together. As one dimension is stressed outside its normal resonance, **the others MUST follow suit.**

Particle Resonance

I need to drive this stuff home a little and to do that, let's look at particles.

The difference between a helium atom and a gold atom is vibration, but what keeps the gold atom together? The answer is resonance.

Just like the electro-magnetic resonance, particles express this same feature. Particle resonance is the most comfortable frequency for the Aether [potential for having mass] and gravitational field to stay. It is a point where Aether and gravitational fields both have the same strength and when that occurs, the affect of the 2 fields is most stable. If these two fields are stable at a high frequency, they appear to be a large atomic mass. At lower frequency resonances, the lighter atoms become apparent.

You might have gathered from the similarities of the various examples that the esoteric components of life and consciousness would also have this resonance feature and its manipulation can be described and so it does. In this case, a life force and consciousness level would be matched to provide the most stable life pattern.

At Speed of Light No One Ages

Let me try something a little different in describing life resonance by using light. While light is constant, we have also determined that there is no aging if someone is going the speed of light to us. It makes no sense to us, but we sort of accept it. Atomic clocks being sent in space ships come home showing the "experienced time" in the space ship was reduced. From these experiments and others, there is little doubt that life is, somehow, suspended at high velocities. Unfortunately, it is not a simple observation. If someone was going to the nearest star at very close to the speed of light, he would get there in about 4.5 years. We could watch the event and record the event, but the man in the ship would still not age. His universe slows down as he speeds up; to him almost no time passes. If he shines a flashlight about halfway to his destination, the light from the flashlight would get to the destination before he got there in his world, but about the same time for all other viewers.

This reduction in aging is the best evidence that the Life dimension is associated with individual universes or associated with universes linked by a common velocity. Remember, it doesn't matter what direction you move to cause this affect. If you were simply **vibrating** at the close to the speed of light, you would age very slowly and everything around you would simply start rotting before your eyes.

Carnal and Spiritual Life

All that aging stuff is carnal living, but some aspects of living are not carnal, they are spiritual, and sometimes the spiritual aspects of life become very pronounced. They don't follow what we normally believe to be truth, but that does not mean they are not real. People really can bring others back to life, turn water into wine, walk on water, and lift up cars to protect those they love.

What is Life?

I think you have probable wondered about this at one time or another and you have never gotten a straight answer. I know you have heard about someone climbing to the top of a Tibetan mountain to talk to some Guru about this subject only to find that his answer was no answer at all. Well! I'm going to try to relieve some of that mystery and flim-flam.

The answer isn't simple nor can it be straight forward. Before you can see the truth, some relearning is in order. Life is not an embryo expanding to become an animal or human and it is not the combination of DNA structure to form a plant of germ. Life isn't controlled by particles or electro-magnetism. We had to establish an entirely new set of dimensional qualities to even attempt to look at life. Unfortunately, for those who wish to see our universe devoid of a Creator and see it as a machine continually building, destroying and building again, you are going to have problems. There are some who try to define life and death with chemicals and sperm, but the ancient book of "Secrets" found among the Dead Sea Scrolls may help us in our desire to determine the truth.

Book of Secrets

The "Book of Secrets" found with the Dead Sea Scrolls provides a strong warning about the use of "secrets of God". The main secret implied was the secret of life. It says that because we, as humans, don't understand what we are doing as we manipulate "Nature" it will only end badly if we try to build "Life". Of the secret elements indicated in the text, it seems that the "manipulation and attempts at creation of life" is the worst on to try. Here is a small excerpt of the book. Its message is repeated over and over in many of the ancient texts. (Portions in parentheses were unintelligible.)

*"--With (this I beseech your attention. All of the) secrets of sin (and life were attempted) but they **[preflood humans]** did not know the secret of the way things are nor did they understand the things of old. They did not know what would come upon them, so they did not rescue themselves **[as smart as the preflood people were most died anyway.]** without the secret of the way things are. ----(God controls) every secret, and he limits every deed and what (magic that is known by) the Gentiles **[People that were not descended in the line of Noah]** , for He created them and their deeds also ---You have not become wise in understanding (my secrets); for you have not properly understood the origin of Wisdom."*

The book is saying that God warned us against trying to create life.---It can't be done because no one knows what life is. At best, we can understand the self portion of life, but trying to see the significance of both the Soul and Spirit become confusing. Possibly, we could build a life that could be infiltrated with a "wandering" soul and the thing would be similar to life. With that let me discuss Clean and abominable animals described in the Biblical texts. Most of the animals today were, evidently made by people from before and immediately following Noah's flood. A list of animals is provided to us, but it sounds so bizarre. For instance, all apes, whales, porpoises, eagles, reptiles are abominations leaving locusts, and sheep and others to be the only animals God sanctioned as not being abominations.

How Can They Live?

Potentially animals could live without souls, but a more reasonable possibility is that life has some elements of soul or they would not exist. The souls of those who died without the "light" were cursed to stay wandering on earth without a link to reality unless they could "find" a host. Jesus, in the New Testament allowed the "demons" to enter pigs because they "normally" went into pigs rather than be completely miserable in"nothingness. These abominations that were "Created" and described in the "Book of Secrets" most likely, became hosts for many of the demons who ones controlled the world.

The Sumerian texts tell us that ancient humans "created" huge animals to fight against God in the ancient wars. Jewish, Greek, Egyptian, and Sumerian texts talk about misshapen people and animals that were the results of experiments with life. The Bible indicates that most of the animals that live today were ABOMINATIONS. We certainly can assume that these abominations were attempts to create life by modification of genetics, and other inappropriate things. If we try to understand life by trying to manipulate DNA and build new species, we will suffer for the blatant disregard for God's warnings. That being said, I don't think the quest for understanding what life is should be considered in this warning. After all, God made us inquisitive.

Anthropic physics must be introduced to you and reality as you know it needs to be completely redefined.

> **The ONLY way to define Life and death is by looking beyond the incorrect, generally limited, unfounded, and carelessly defined 3-dimensions we have been taught in school.**

Many of the present theorists know that this is the only self sustaining theory that has not been smashed into bits. String theories, black holes, big bang, super symmetry, tamashii modeling of mass, and all the rest can't be defined completely without external universes and control of this one by the unseen expanse we can call the "soul collective". That is because all dimensional qualities of this "seen universe" are in direct opposition or resonate with a collocated, unseen universe some call Heaven, some call universe B, and some call linked universe. Two things hold the universes or the second half of this universe those are this collective and time reversal.

Life Conversion

It seems that everything converts to other useable characteristics in our linked universes. As out-waves [mass vibrations] push their way to the limits of our universe, they are characteristically converted to in-waves [force vibrations] used to apply stresses on our linked universe. As out-waves emanate from our linked universe into ours, they are converted to in-waves to apply forces to our out-waves. That whole machine, while exotic, sort of, makes sense by now, I hope. I know the quantum mechanics dilemma threw some of you as time space is not a major definer of vibration, but that was mostly because our concept of vibration is through space and that is not what it is, exactly.

> **More importantly, life must have the same conversive effect in our joined universes. It's bigger than being born, living and dying.**

It's more expansive than "I think, therefore I am!", as Descartes used to say when trying to define life. It is more noteworthy than "Life is a box of chocolates." It is more complex that "I die and go to heaven." It's more complex than light and it's more wonderful than the birth of a new baby. Here is what a modern "seer" had to say.

Edgar Cayce on Light and Life

Edgar Cayce, the sleeping prophet of the twentieth century may give us a little more perspective.

"In the manifestation of all power, force, motion, <u>vibration</u>, that which impels, that which detracts, is in its essence of one force, one source, in its elemental form. As to what has been done or accomplished by or through the activity of entities that have been delegated powers in activity, is another story. [Of course, the one source he is talking about here is VIBRATIONs not recognized in this universe.]

"God, the first cause, the first principle, the first movement, IS! That's the beginning! That is, that was, that ever will be! The following of those sources, forces, activities that are in accord with the Creative Force or first cause - its laws, then - is to be one with the source, or equal with yet separate from that first cause. When, then, may man - as an element, an entity, a separate being manifested in material life and form - be aware or conscious of the moving of that first cause within his environment? Or, taking man in his present position or consciousness, how or when may he be aware of that first cause moving within his realm of consciousness?" [Cayce is indicating here that the first cause vibration is the thing we must be conscious of if we are to understand life.]

"In the beginning there was the force of attraction and the force that repelled. [Dr. Milo Wolff would have said IN-waves leave the universe and Out-waves enter the universe. We might see that forward and backward time must be similar]

Hence, in man's consciousness he becomes aware of what is known as the atomic or cellular form of movement about which there becomes nebulous activity. And this is the lowest form (as man would designate) that's in active forces in his experience. [This is the basic Carnal Life haplessly enjoyed by most.]

Yet this very movement that separates the forces in atomic influence is the first cause, or the manifestation of that called God in the material plane! Then, as it gathers of positive-negative forces in their activity, whether it be of one element or realm or another, it becomes magnified in its force or sources through the universe. Hence we find worlds, suns, stars, nebulae, and whole solar systems, moving from a first cause. [These would be a parallel universe filled with the same things we find in the "Carnally linked world". The description of God being the first cause speaks to the other universe linked to spiritual life.]

"When this first cause comes into man's experience in the present realm he becomes confused, in that he appears to have an influence upon this force or power in directing same. Certainly! Much, though, in the manner as the reflection of light in a mirror. For, it is only reflected force that man may have upon those forces that show themselves in the activities, in whatever realm into which man may be delving in the moment - whether of the nebulae, the gaseous, or the elements that have gathered together in their activity throughout that <u>man has chosen to call time or space</u>. And becomes, in its very movement, of that of which the first cause takes thought in a finite existence or consciousness. [Finite existence is the carnal element of the Universe and its negative time characteristics are certainly described as a mirror or mirror image of our universe.]

"Hence, as man himself applies himself - or uses that of which he becomes conscious in the realm of activity, and gives or places the credit (as would be called) in man's consciousness in the correct sphere or realm he becomes conscious of that union of force with the infinite with the finite force. [This is speaking to the possibility of changing ones "resonance" by acceptance of the

spiritual universe and its implications. Finite is carnal, infinite is the infinite spiritual life.]

"Hence, in the fruits of that - as is given oft, as the fruits of the spirit - does man become aware of the infinite penetrating, or interpenetrating the activities of all forces of matter, or that which is a manifestation of the realm of the infinite into finite - and the finite becomes conscious of same. [Going from the realm of the finite to the realm of the infinite speaks of becoming a free soul through change in resonance or complete separation from the Carnal body in death.]

"It may be said that, as the man makes in self - through the ability given for man in his activity in a material plane - the will - one with the laws of creative influence, we begin with: Like begets like - As he sows, so shall he reap - **As the "man thinks in the heart, so is he."** [As Jesus said, faith of a grain of mustered seed could allow moving mountains with our minds.]

"These are all but trite sayings to most of us, even to the thinking man; but should the mind of an individual (the finite mind) turn within his own being for the law pertaining to these trite sayings, until the understanding arises, then there is the consciousness in the finite of the **infinite moving upon and in the inner self. Life in all its force begins in the earth as the moving of the infinite upon the negative force or the finite in the material, or to become a manifested force**.*"* [We can stay in a carnal world or enjoy the fruits of that experience that is beyond.]

To Dr. Casey and the insights that he obtained from his visions, or whatever they were, life was an extension of existence. Not only could life modify existence, life was and is existence. Without consciousness, Einstein told us there was no reason to define existence. It goes farther than the "Positive Thinking" described in a number of books where believing that you are rich and miraculously you will become rich. It is more like the entire universe you perceive is slightly different than all other universes perceived by other people because they are defined by one's life resonance.

A Conscious person not only can control his destiny, but also manipulate the world perceived by others around him or her.

Welcome to the Ethereal Dimensions, Adjacent universes, and our Ekpyrotic membrane. The Ekpyrotic thing is also called the "M theory", and it is the one that makes sure there is a conservation of dimensions between our linked universes. Actually, the re is a conservation of everything, including Life/death.

Conservation of Everything

Here is the main problem. We now know that everything is vibrating from energy nodes [or fermions]. The vibrations NEVER stop. They continue to move away from the central core. Just as Einstein had theorized, soon, they would be lost at the end of the universe [whatever that is.] We would run out of particles, photons, and energy. **We would run out of life itself.** Luckily, we find that there is a conservation of everything that sustains the universe.

Conserve Energy-It's the Law

Let's first consider the common "law" of conservation of energy. All we can do is to change the type of energy we cannot add to it or take away from it. In recent years, some have identified areas that seem to create matter which suggests the creation of energy. If we create matter, certainly, we create energy and we create something else that is very important. What we will find out that conservation of energy is not a true concept, but symmetry of energy absolutely works and I'll show how it affects life. Not only is there symmetry of energy defined in our universe, but also the following:

- **Symmetry of Energy** While Einstein indicated that it could not, by itself be conserved, we find that by a new theory called Super Symmetry; energy can APPEAR to be conserved.
- **Symmetry of Matter**-As a black hole produces matter, there is no issue as anti-matter is reduced at the same time. I'll have to make anti-matter more than a comical word for you on this one.
- **Symmetry of Light**- Einstein was always worried that as these photon things approached the end of universe soon, the light would leave our universe and the number of photons would continuously be reduced. We will see that Light is sort

of rejuvenated by anti-photons, but that by itself does not allow us to appreciate the electromagnetic vibrations. That required a little ethereal physics.
- **Symmetry of Life-** As people die others are born. As people become more spiritual, others become less spiritual. If someone's life is ended here, it is started in another universe and vice-versa. I know this sounds like yen and yang but it is more than good and perceived evil.
- **Symmetry of Death-** As people die they must be born or their consciousness does not continue. If it does not continue, the reality established by that consciousness ceases to exist and reality will become more and more skewed. Soon physical laws would have to be changed to support new life that had not been established from a previous life. While this is not a true symmetric requirement, we will find later that death must be recycled just like life.

Everything in the universe is regenerative. We simply need to broaden our horizon to find what is feeding the dimensional elements of the universe

This one is part of the answer to that Life/Death Question. The thing that is continually controlled is the Collective Soul.

Conscious Collective

Another way of looking at this ethereal component of the universe has been ADRESSED is to call it the conscious or soul collective, but what one really needs to understand to fill in some of the blanks was presented by Einstein many years ago now. That idea was that relativity is not simply how space and time are modified to present a common image to an observer, but that –

> **EVERYTHING seems to be modified DIRECTLY by this conscious collective thing-a-mo-bob. Don't go saying "We are Borg" like the Star Trek Nemesis, but in a way, we establish a unified reality.**

I'm going to get you comfortable with this whole ethereal dimensional model and the rest is up to you. Once you understand how it all gets put together, you will have to build on these kernels.

> **In order for life to exist and continue, there must be the rejuvenation of life that comes as in-waves from outside our universe or the component never left. The soul sticks around and the spirit leaves.**

God Took the Tree of Knowledge from Man

Life combines self, spirit, and soul to the universe in a special and important way. Once we see the significance of our particular lives, we can begin to understand how much power we posses in this universe. There was a reason why God pulled the Tree of Knowledge away from people. Whether the description was figurative or descriptive, one truth is presented in the Book of Genesis. If man knew the innate power presented in his creation of conscious life, we would mess up this world and limit interaction with our creator. On the other hand, knowledge can be split into two confines—carnal knowledge and spiritual knowledge. <u>As one expands the carnal nature, there is a loss in "communication" with our adjacent universal duals. Expansion in a spiritual way, changes our resonance which modifies the resonance of the perceived world towards a unified communion with our adjacent universe. Everything is interactive, so to speak</u>.

Vibrational Distinction

In the operational dynamo, what we call light frequencies become radio waves as frequencies are decreased. In the diagram below the vibrational frequencies of matter are in the exahertz range [exahertz means "quintillion cycles per second"]. The frequencies have been converted to the distance traveled during one cycle at the speed of light so don't go out and try to whistle some gold in your pocket. Gold vibrates at 60 exahertz meaning it wavelength is 5×10^{-6} microns. That's really, really fast.

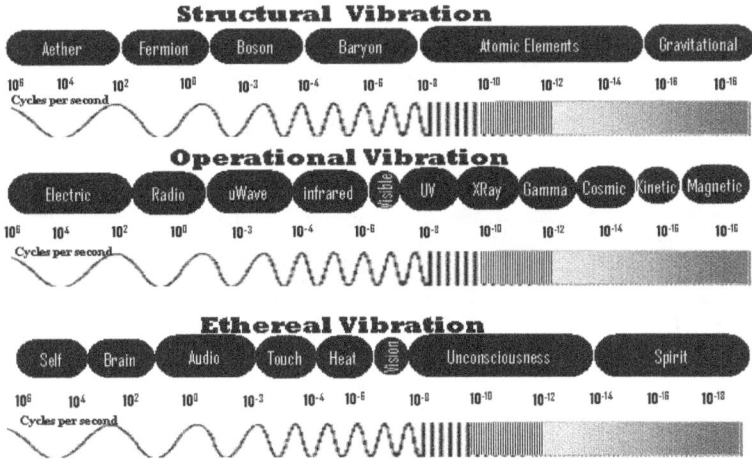

Quick Definition of Ethereal Spectrum

The main thing I want to introduce by this chart is the ethereal vibration spectrum. Notice that low frequencies can be described as self or carnal levels. As the vibrational aspect of this dynamo increases, the Spiritual component or sub-consciousness takes control and finally spiritual pureness is even at a higher level. Low levels are associated with survival, sex, and hunger. The lowest of these levels is life without true consciousness like a snail. Its life consists of survival with a very few ventures into sexual gratification and nothing more. One could say its life has potential, but no real force. The higher frequencies are associated with enlightenment and finally spiritual awareness. Guess what! If you become purely spiritual, you cease to exist in this world. All living entities stay somewhere in between. The noticeable characteristic of Life is that it has a resonance. The more spiritual one becomes, the <u>less carnal he must become</u>. While this seems sort of obvious in a life description, the amazing thing is that the other 2 vibrational spectrums do EXACTLY the same thing. This begins to confirm that we are on the right track. Also, understand that if we want to change our spiritual level, we must reduce our carnal level or our vibrational level must be increased. The question might be, "Just how can I get the vibration level higher?" and what are the Life "in-waves" alluded to By Dr. Milo Wolffe?

What Is A Life In-Wave?

I've been kicking around these vibration "standing waves/nodes" and these odd in and out waves, because some physicist identified them that way.

I'm sorry!!

Let's try to put a perspective on them. The out-waves are fairly easy to understand as vibrating nothings emanating from each of these standing wave point. OK! Not very easy, but at least one can generally understand the concept. Life out-waves are generated by conscious understanding of the environment. [Self, Survival, and Sex]. None of these attributes allow modification of the self resonance and all can be identified as CARNAL or generated from inside the universe creating out-waves that spread out to the limits of the universe. [The body is inconsequential]. OK it's nice to have a body, but what I mean is that the Self, Survival, Sex life of a bug or germ or even a person doesn't affect the other dynamos. They are Static and generally identified as "the self dimension". The out-waves keep going away from the "node" to infinity and --------beyond [never to return???????]. It would seem that life would vanish from the universe over time.

Life Doesn't Vanish

For life, out-waves constitute the carnal aberration of living. All aspects of self can be easily shown to have initiated in the universe and as they try to expand outward to our adjacent universe, then something magical happens. They are bombarded with in-waves of life just like particles were bombarded to establish stress in particles so that work can be done in the universe. This bombardment alone should allow us to know that other beings are resident in the adjacent universe. Their survival dimensional vibrations continuously leave and are converted to in-waves that CAN react with ours if we want them to.

Some More from Milo & Albert

To try and get a feeling about what in-waves are, Einstein and Dr. Milo Wolffe will help. Dr. Wolff stated, *"Forces/Fields are caused by wave interactions of the Spherical In and Out Waves with other matter in the universe which change the location of the Wave-Center and which we 'see' as a 'force accelerating a particle."*

In English, this means that these in-waves make force to establish energy needed to run our universe. Unfortunately, there are a couple of real issues. Where do these all important in-waves come from and how do in-waves relate to living?????

These In-Waves Are Different

As Carnal life spews out of this universe, we can assume it can be a driving force in our linked universe that has living substances spewing out Spiritual life that regenerates our Carnal living.

Life's Out-Waves

You may know what I'm going to say, but I will say it anyway. The in-waves are coming from an adjacent universe. When they come into out universe, they act backward to the out-waves produced as various out-waves come in contact with each other to produce vibrational nodes we describe as atoms or atomic clouds. <u>Because the in-waves are backwards, contact with out-waves cause STRESS we call conscience.</u>

Think of it this way. Carnal out-waves stressing self/survival/sex emanate outward and they are countered by in-waves that stress the exact opposite which are true love, true concern or empathy, True willingness to die for another, true spiritual insight. The next time you hear a little voice telling you not to be selfish, or eat some food you should not have, or giving you some super strength to save a life, understand what that really is. It is simply the In-waves of life just like we have been going over and over, as they are represented in the other dynamos. One can think of them as backward selfishness, or backward hate.

The in-waves are not backward to the adjacent universe. Time is reversed to assure conservation of time and life is reversed in the same way. In the adjacent world, the Self dimension IS the Spiritual awareness/empathy, and love. The out-waves from our universe are mostly self/sex/and survival which are used by them to allow them to understand themselves in some way. If that isn't odd enough, as time goes forward here, the concept of time in our neighbor MUST BE backwards to us but forward to them.

Because the in-waves are in opposition to all the in-waves, when they come in contact, they put stresses on the life's vibrational nodes. That is what we think of as <u>carnal stress</u>. Carnal stress has limited effect on our lives unless we can change our life resonance. That is coming up but it may start sounding religious so I'm just warning you now.

Life Would Disappear
If we had no life energy inserted from out adjacent universe, life would simply have no meaning at all.

If someone tried to inform you of one basic truth about life, it would be that living without a heaven is simply existing and life in our universe would eventually disappear.

So what! You created a great food dish or ran a mile, or saw a sunset, or became an evil dictator. The motions of life would simply have no meaning, they would be actions controlled simply by the environment. The word predisposition would be will rather than could be. There would be no moral or immoral. There would be no good or bad. There would be no "substantial" happiness or sadness. I say substantial here because humans and robots, for that matter, can fool themselves into thinking they were happy or sad or depressed or destitute. Possibly even a germ thinks he is having fun.

The ancient Jewish religious books and our Bible indicated that God took the light away from Satan and his followers and Jesus told his followers that he was the "light of the world". While there is some reason to classify these things with the study of light, another more reasonable way to view these statements is with respect to life. These statements could have been said a different way.

God allowed the life resonance of Satan and his followers [From the adjacent universe] to remain Carnal giving them little reason to live.

Jesus To the Rescue

Besides coming to this universe and dying for our inability to accept anything from the spiritual in-waves, Jesus [incarnate GOD/Creator] brought a new way to expand our resonance and allow spiritual in-waves to expand, rejuvenate and give meaning to out lives. Certainly, Jesus' resonance gift is not the only way to change one's life, but it seems to be a good one.

Oh, Yes! Jesus' gift had a name [Holy Ghost].

I know it's a weird name for a gift that would allow us to modify our resonance and allow interaction with our adjacent universe while we are in a carnal body and even afterwards, but the name is just a name.

In and Out Wave Differences

The reason out and in ways are opposite is that adjoining universes have "backward time" to each other. I know I've stated this a couple of times, but it needs to be understood. As an out-wave leave our universe, it appears to reverse by the time dimension of the outside universe and it is turned into an in-wave. Guess what!! The out-waves in that universe have vibrations in opposition with the newly created in-waves. As these out-waves leave the adjacent universe they enter our universe and appear to be opposite because of our backward time to the other universe. Of course what that means is particles in this universe become forces in the adjacent universe.

When studying life, one must also sense the difference in experience. Out-wave "carnal" experience is substantially different than "Spiritual [in-wave] experience. Both must be levied together to allow for a more meaningful life experience.

God knew that Carnal existence and the feelings of self, sex, and survival were too strong so one of the three dimensional components of GOD was sort of introduced to channel the spiritual in-waves into our carnal selves.

Not Sacrilege

I know some are thinking this talk is sacrilegious and others are saying that I should not be confusing science with God and the third group is saying that this mumbo-jumbo means absolutely nothing. Well, I see everyone's points, but the facts are as indicated below.

- God's trinity was described that way for a reason.

- The Holy Ghost being introduced into our world to allow us to "accept" a gift from God, sounds pretty much like what I have been saying and it is remarkably similar to the interactions of the Structural and Operational dynamos.
- There is always a struggle in ourselves to do the right thing followed by doing the most comfortable thing.
- There is an innate carnalness to mankind, no matter what we say.
- We know that consciousness does affect reality. It has been somewhat proven in experimentation and substantially in calculation.
- If life did not have a way to be rejuvenated, soon there would be no life in the universe just like there would be not light and all matter would be in a state of entropy.
- God being a trinity only makes sense if everything else in the universe is made the same way and dimensional characteristics are in groups of mutually perpendicular threes.
- The power of positive thinking and self actualization absolutely do change our environment just as written about in dozens of books.
- Life is more than chromosome multiples in a body. That is simply stupid.
- Life is more than chemical reactions in a brain. Just like sight and vision are completely different, life and predispositions are completely different.
- The will to live is a very strong force that has nothing to do with the brain.
- The conflict of once thoughts to do good and/or something that will satisfy self is not caused by chemical reactions in the brain.
- When your body dies, life does not. It either stays in this universe or it goes to the adjacent universe.
- Some methods were discussed in the preceding book to allow travel into the adjacent universe to provide a mode for time travel.

Let's see how we can go to the next universe.

Going to Another Universe

Particle Transfer

Let me take this one more step. Let's say a person from our adjacent universe goes faster than the speed of negative timed light in his universe. He would immediately be transferred to this universe and become force rather than matter. I know you are thinking about how watchers/angels seem to be able to make this transition and can still be touched and are just like people so there is something else going on with life or the watchers wouldn't be showing up. This type of transfer could possibly be done without changing the resonance of life and development of a more spiritual lifestyle, but there would certainly be issues in understanding a world where their perceptions would be different that the perceptions of those around him. If a carnally minded person lived in the heaven universe, he would always feel completely out of place, out of sync, and unable to assimilate. When the ancient texts talk about how Satan and his followers had their LIGHT removed, possibly it was talking about this difference in mindset or vibrational level.

Where Is This Universe?

Don't go thinking that our linked universe is co-resident in our same space just because watchers appear and seem to disappear instantly from this heaven place. There has been "NO" credible evidence that our linked universe [and perhaps more than one] MUST be connected physically. The whole concept of quantum mechanics also gives us a better understanding that what we perceive as placement in time-space has little to do with reality. The time-space reality we enjoy is held together by common understanding of the environment we live in. Don't worry about everything you know today changing in a blink of the eye just because others decided not to think that way tomorrow. It doesn't work like that. Changing the group consciousness is not something that can be accomplished easily but there are some glimpses to it being done all the time.

- When someone lifts a car to save a child, it can't be done except in the modified consciousness area. There must have been a very strong yearning to have the slight change in reality and the creator also might have stepped in to allow for the change as well.
- When Peter walked on water, he certainly was able to change his environment by changing his "life's resonance" but he also could have been joined by his fellow shipmates and don't discount Jesus, God incarnate, being around.
- When various monks were able to lift off the ground and fly, they had to stay in a state of meditation and prayer for some time before this little change in environment could be accomplished.
- Faith healing, or what has been termed that, surely shows a shift in our normal reality and a change in the common

consciousness. Part of this is accomplished by convincing everyone that it can be done, convincing one's self that is has been done, and probably
- These guru characters that bury themselves for days without enough air, food, or warmth to assure survival, and they return unharmed must have shifted their consciousnesses.
- When Keely's experiments showed such promise and then fell apart as others tried them until he placed his hand on them, we can assume that something miraculous was happening. He, somehow, affected his life resonance and that affected the near term reliability.

Life and the Out-wave

Just like matter has no end as Einstein and Wolff so eloquently described, life has the same characteristic. There is no specific end of consciousness. Some might say that this statement is the ravings of someone who simply cannot fathom an end of his or her life. Others would take a more religious approach and say that life ends and life in heaven or hell begins. Rather than trying to understand the oscillation characteristics of life's ebb and flow or the renewing of matter as it reaches the ends of our universe, one might better understand the concepts as symmetry existence rather than conservation of mass, energy and life. These things are more cyclic than constant.

Symmetry Not Conservation

As I mentioned slightly before, symmetry is more important and more realistic than conservation of universal elements. You may know about the theory of Conservation of Energy. Energy simply changes state rather than dissipates. It has an endless oscillation or vibration between Static and Kinetic elements. In a way, this is a true statement and certainly we can look at our universe in this way and things seem to fit together. What if I were to tell you that static energy, or what we call static energy is a characteristic of the out-waves from our universe and kinetic energy is a characteristic of in-waves from an outside universe. We don't conserve energy. It is continuously replenished. As our universe outputs out-waves, they are converted to in-waves and returned.

The universe dual MUST run symmetrically. The only way to apply force in this universe is from an outside force. AND Guess what!!! All dimensional qualities of this universe MUST BE symmetric with its universe dual.

Oh, you are a smart one!!! You are thinking, "That's not symmetry, the examples I keep bringing up show the opposite to symmetry.

Got You With This One!!!!
Remember that our linked universe has backward time. Therefore decreasing matter backward in time is exactly expanding matter in their time. Both universes would experience expanding masses as matter is created over here, anti matter is increased over there in

reverse time. [My head hurts!]

Backward Living

I've opened up a bag of worms now. If both universes sense time backwards to each other, there could never be interchange between universes------Right????-------

Wrong!!!!

In our linked universe, people experience time differently. It would be hard for us to understand how they see time, but I do think there is exchange between our universes so this oddness simply must be. I think that explanation will take one of those gurus on top of a mountain to explain. I'm having a hard time with forward time and I think I can still explain life without the extra confusion

Certainly there could be more universes, The Membrane [M-theory] suggests many more universes could be co-resident with our "visible one" and so do many of the string theories and the symmetry of our universe may be shared, but it could also be indelibly linked and presented in this work.

What I was saying is this. Matter, Energy, Life, Time and everything else in this universe stays in constant motion. Never increasing and never decreasing without a secondary outside force. Let me give some examples-

- If **kinetic energy** decreases, it increases in our joined universe which in-turn causes our STATIC Energy to Increase and vice-versa. [If we make a conversion of energy in our universe, the opposite WILL occur in the joined universe.]
- If **Gravitational energy** decreases, it increases in our joined universe which in-turn causes the exact opposite to increase in our universe. This increase makes us have an apparent reduction in mass so we will have to look at it a little. [If we force a reduction in gravity, our joined universe WILL experience an increase in magnetism to compensate.]

- If **Photonic [Electro-magnetic] energy** increases, it increases in our joined universe which in-turn causes the exact opposite to decrease in our universe [Because in-waves turn into out-waves, electromagnetism become particles in the adjacent universe so mass would reduce.]
- If **Internal Life energy** decreases, it increases in our joined universe which in-turn causes our External Life Energy to Increase and vice-versa. I know this whole external energy sounds weird, so we will look at that a little as well. [The main thing is that as internal life here increases, it WILL decrease over there.]
- As Time vector goes in one direction, it MUST go in the exact opposite direction on a joined universe. **[Time must go backwards. This, of course is from our point of view.]**

Certainly, these may not be the exact duals of our universe, but hopefully, it will give you an idea about how each one of the elemental parts connect with both internal dual dimensions and external anti-dual dimensions.

To carry it one step further, we can establish that is the energy bond for the Particle dynamo is [Static and centripetal] forces and the operational dynamo is associated with Kinetic and centrifugal forces, the opposite would be true in the joined universe. The anti-particle dynamo would be associated with Kinetic energy and the anti-force Dynamo would be associated with Static energy in that world. All this stuff is good information, but you really want to learn some more about how to attain a higher level of awareness, better life, and death that are more meaningful. While all this symmetry is going one we still need to remember disconnected souls.

Ghosts

Ghosts or demons, have a life essence that cannot leave our universe. When a change occurred from this living entity to a dead one, something very strange had to occur to balance out the universe. When they died, they did not return to life. Oops! I'm sorry I said that before you were ready. Conservation Science tells us before one can have his consciousness and living body "Die" an equal measure of consciousness and life must be regenerated or he CANNOT DIE.

You might ask, "What about these angels that appeared out of nowhere? How does conservation of energy and matter account for them?" Many respectable people have seen angels from time to time who, apparently, leave their nonphysical realm to appear here in ours. If they can interact with our material environment, they must be at least partially composed of matter themselves since it is an observable fact that only matter produces the radiation, gravity, and mechanical forces that affects other matter. By disappearing in their "heaven world" and appearing in ours, many "scientists" indicate that they violate conservation of matter and energy in both worlds! "This simply cannot occur", the Scientist exclaims! For the strict energy conservationist, the whole "change from life to death or ANY human" gives him heartburn because energy seems to be lost so they simply called it an anomaly.

Your life energy simply cannot cease to exist. I know you are just saying---so what, another person simply is created and uses my life energy when I die.

That sort of is correct in that new people, personalities are created ALMOST every time someone dies. Fortunately or unfortunately, there must be also a limbo place where consciousnesses react in the Heaven or our universe. Hopefully, I can explain the physics of all of this to you in an understandable way and tell you about some very unique evidence that is being established by regression scientists. One in particular is a man named Michael Newton. I think you will love the details that he has uncovered.

Michael Newton's Soul Journey

Michael is a hypnotherapist in California and he is good at it. His specialty is regression therapy. As many others have found, sometimes, fears or limitation we have in this life come from events that occurred in a previous life. Too many people have been regressed for it to not be considered a reasonable method of research and it cures people. Anyway!!! Michael started getting information from his patients that didn't make sense. They were talking about people they had contact with BETWEEN LIVES. Case study after case study, more and more data was collected and confirmed by the details present by others until a reasonably clear picture of this purgatory type place you have heard about and dismissed. When events of a life are not completed in a reasonable way or the person feels anguish at something that happened or he did while alive, he relives those events on the other side to experience the other side and gain empathy. I'll bet you thought you got empathy because of your marvelous insight to your surroundings. Well!!! There is a growing amount of data that suggests that empathy comes from ACTUALLY living a life as the OTHER individuals. Thomas Maslow's Self Actualization level that forces the insight concerning those around us, may not be a carnally learned event, but is accomplished over many existences.

Helpers
Here is the other thing Michael found. There are Helpers in this purgatory place that help us understand what our previous weaknesses were. There are also un-embodied friends or acquaintances and groups that sort of stick together in a journey

that continues for eons. I would not be surprised that some of these "helpers" are the original rebel's from the Heaven Wars. What else are they to do?

Is This Sacrilege?

Wow!! This sounds so absurd, you must be thinking, I had better not go outside if lightning is in the air. God will strike me down on the spot!! Let's review some of the Biblical statements again and get back on track.

Judas was told it would have been better for him not to have been born, even though he repented just like all of the great followers and he had so much remorse that he ended up taking his own life. If Judas went to this place Michael found after death, it all makes sense. God would want him to learn, get punished, regenerate and gain another chance at taking the "Light" of the Holy Spirit" in a new body. If the next didn't work, anther would be possible until finally a larger percentage of the population would be able to accept God before his final return. Therefore, Jesus would not return for thousands of years to give as much chance as possible.

God does all things for good- according to New Testament quotes. Deadly hurricanes, horrible sicknesses leading to death, and a wide assortment of troubles are for your good. This makes much more sense if your death is only part of your life experience and should who had not gained the Holy Spirit "Light" are allowed to try again in a succeeding life.

Reincarnation taught in the Bible now makes sense. Why would there be reincarnation unless there were new things that had to be thought to the reincarnated souls.

Samuel- being summoned by the Witch of Endor from "purgatory makes sense and his being conscious makes more sense. Also, his ability to see the future makes more sense as we will investigate in a little bit.

Reanimation-The idea that when Jesus returns, the dead in Christ rise up out of their graves first as attested to in a large number of Old and New Testament scriptures, makes sense as there is somewhere to rise from that is not just dirt.

Rebels- This whole concept presented in the Bible and many other texts about the losers of the Heaven Wars being dead, but never being able to leave the Earth makes more sense.

More Reading

Michael's books, "journey of Souls" and Destiny of Souls" provide a very good overview of this new science of intra-life regression so don't just take my word. Also, read "Many Masters" by Brian L. Weiss M.D. or contact a growing list of others around the world. Below are a number of Life between life therapists experimenting in this exciting new area where we can gain more understanding of death---while we are alive.

- **Australia-** Tony Collins, Tania Dionisio, Tracey Robins
- **Czech Republic-** Bernadeta Hodkova
- **France-** Kathy Gibbons and Bernadeta Hodkova
- **India-** Neeta Sharma, Blossom Furtado, Asis Ganguli , Jyotika Chhibber
- **Ireland-** Kathy Gibbons, Izabela Fouere
- **Norway-** Lise Stjernholm, Tone Hansen, Lisbeth Lyngaas, Kjus Garden, Helen Soernes, Bodil Rosvik, Alice Kjolerbakken
- **Singapore-**Reena Kumarasingham, Peter Mack, Antoinette Biehlmeier
- **Sweden-**Lis Lindahl, Karin Danneker
- **Switzerland-** Radovanovic Kchler, Jacqueline Niggli
- **UK** -Hazel Newton, Liz Kozlowski, Elen Clulow , Reena Kumarasingham, Ian Lawton, Peter Blayney, Maggie Amuna, Janet Trelaor, Anjalee Carey, Lorraine Flaherty, Wissam Awad, Reena Kumarasingham, Doug Buckingham, Tricia Allen, Bridget Rattigan , Liz Kozlowski, Katherine Membery,

John Nicol, Chris Hanson, Dave Graham, Debbie Wild, Linda Hopkins, Janet Treloar, Trish Heenan

It looks like if you visit England, you can find out a lot about what this book is about. I wonder how many hypnotherapists DON'T try to determine what you did between lives.

I know this has been quite a bit to spring on you so let's start expanding on some of the more important themes before going into why some of this HAS to be or Physics would not work. Our first stop is the brain.

Brain Classification

So many people base life on brain function, I think that we need to look at the thing for a minute and sort of expand on the "normal" definitions we have come to love. While the brain activity doesn't directly explain life, it may provide another level of insight. Let's investigate further. The brain functions are, many times, segregated into 5 parts; Epsilon, Delta, Beta, Theta, Alpha, and Gamma. While vibrational characterizations are not clearly understood, many try to test and experiment with the various brain frequencies just to see what happens.

Epsilon Brains

Epsilon or vibrations less than 1/2 hertz are not very normal, but there are some who can sort of suspend their brain functions and report **spiritual insight and ecstasy** as if it were an out of body experience.

Delta Brains

This is sometimes called the **consciousness of survival**. Delta or vibrations less than 1/2 to 4 hertz seem to be a **remedy for anger** and causes **confusion, disorientation, lucid dreaming, and a decreased awareness of the world**. One might experience this brain function vibration level if he were in a **trance** or under hypnosis.

- **Medulla Oblongata-** As one would imagine, this brain component executes the most important functions of survival including regulating our life processes such as breathing, maintaining a steady heart rate and blood pressure, urination, and defecation.

- **Cerebellum** -Coordinates and controls voluntary movement, maintains balance and equilibrium while walking, swimming, riding, etc., stores memory for reflex motor acts, coordinates simultaneous subconscious actions, like eating while talking or listening etc.

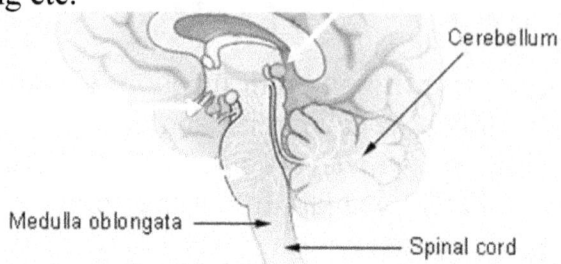

Theta Brains

This is sometimes called the **consciousness of sex**. Theta or brain function vibrations from 4 to 7 Hertz seems to cause <u>**sexual arousal**</u>. It also enhances <u>**memory, focus, creativity and inspiration.**</u> The section of the brain that becomes most active is the pons.
- **Limbic System-** Manages olfactory pathways and functions related to sex, rage, fear; emotions. The pineal gland shown below seems to be sensitive to much higher frequencies that the pituitary gland shown below.
- **Pons-** Manages Respiration. Has control over skin of face, tongue, teeth, expression, and level of arousal.

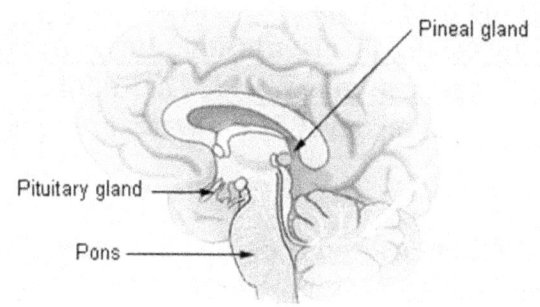

Alpha Brains

This level is sometimes called the **consciousness of self**. Alpha or brain function vibrations from 8 to 12 Hertz seem to <u>**accelerate learning with reducing stress and elevating one's mood.**</u> A person becomes more positive of his situation and may daydream and become relaxed without becoming drowsy. Particular areas of the brain seem to be excited by alpha waves including the following:

- **Cerebral Cortex-** The outermost layer involved in the functions of learning, making decisions, and memory. The alpha and beta brains are established in the back portion. This contains the following:
  - **Parietal Lobe-**Processes sensory input, sensory discrimination.
  - **Occipital Lobe-**Concerned with visual interpretation.
  - **Temporal Lobe-**Section that accommodates auditory performance, speech, and information retrieval.

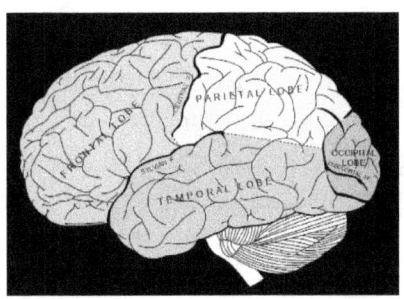

Beta Brains

Also associated with consciousness of self, Beta or brain function vibrations from 12 to 30 Hertz can be induced through active concentration and is associated with alertness, anxious thinking, and analytical problem solving. It also seems to increase judgment or decision making and **survival instinct**. It is as if we **processing information about the world** around us better in this state.

Gamma Brain

This is sometimes called the **consciousness of love**. Gamma or brain function vibrations from 30 to over 100 Hertz really start us going right as our cognitive **skills and focus are heightened**. We have **increased compassion** and perception of reality.

Additionally, we seem to take control over our bodies as **muscle develops, injuries begin to vanish and bones heal**.

- **Frontal Lobe-** Slightly different than the other external lobes shown above, this one establishes memory and cognition. It also enables concentration, judgment, inhibition and personality development. Also associated with the gamma brain are the higher level glands.
  - **Thalamus** –Its main function is providing the brain information on what is happening outside the body.
  - **Hypothalamus** –This section regulates emotions, hunger, thirst, and libido and is responsible for maintaining the daily sleep and awake cycle. To show its strength, it also changes the pituitary gland which changes body homeostasis.

Above Gamma

Little is known above a gamma state except that it appears that the **pineal gland** becomes more active as vibrational ecstasy goes beyond normal gamma levels. We will look at the pineal gland a little more, but just like all the other vibrational dynamos the higher in vibration, the more Kinetic the dimension becomes. The Life dimension dynamo works the same way. Note that all we have discovered to date are very low "Carnal" vibrations emanating from our brains.

Levels of Consciousness

The reason I brought the preceding section was to show that our brains are sensitive to vibration and that what we known about ourselves is basically "Carnal". As the frequency increases, the level of awareness or association with non-carnal life can be sensed. The reason is simple. Consciousness is a vibrational dimension and brainwave studies are not the only way to recognize the vibrational characteristics. A second way is something called chakra so let's look at some of these mystical things. One thing you will notice right away is that there are many levels not addressed in the brain function vibration frequencies normally studied.

According to the believers and the testing skeptics, there are at least seven of these chakras or levels of consciousness.
- Root, Chakra--- "consciousness of food or survival"
- Sacral Chakra --- "Consciousness of Sex"
- Solar Plexus Chakra --- "Consciousness of Self"
- Heart Chakra --- "Consciousness of compassion and love"
- Throat, Chakra --- "Consciousness of the truth"
- Third Eye Chakra --- "Consciousness of our inner being"
- Crown, Chakra ---"Consciousness of the worlds beyond"

They are sort of represented by the diagram below.

I know it sounds like I'm some guru from India talking about chakras, but it is a convenient way to discuss this dimensional component so I'm going to continue. I'm not putting on the towel on my head, but I may hum a little as I write this section.

Survival, Sex, Self

Whether we admit it or not, every one of us battles these things every day. The most basic root chakra is triggered when you get hungry and the sex one, well it shows up from time to time. If you can get past those, you start considering self worth and even love. Many people spend most of the time going back and forth between these two states. It's sort of like a yo-yo. Slow vibrations, higher vibrations, slower vibrations, higher vibrations and still higher vibrations when someone pays me a compliment or a figure out why a light bulb turns on in the refrigerator or I answer one of those "Are You Smarter Than a 5th Grader" questions. Answering one of the "Jeopardy" questions correctly might even get you into the heart chakra, but sex and self still will take control at some time.

Love and Truth

The heart chakra is a vibrational level associated with "Real" love rather than the sexual one. Sometimes this level happens naturally for a brief time and you can't seem to even think about yourself at all. If you work at it, you can get to this level periodically throughout a day and look at people with true look. The Bible called it loving people as you love yourself. Anyway, most just think they get into love and it is more basic. That type of love puts you below the stomach again. Anyway, you must conquer love to some level before you can even get to a point that looks for "real truth" rather than "Vain truth" that we usually accept or desire.

Real truth is a truth that is truth no matter how it affects the event or who thinks it. It is usually not a popular truth or even the one you would hope for. It simply is. While it seems this would be easy to understand and use. People almost never are tuned to this type of consciousness so they accept what they believe rather than what they should believe. Let's say you get an openness to understand real truth, there are still 2 more levels of consciousness to be considered. The next is called the third eye.

The Third Eye

The third eye is derived from the little gland in the brain I mentioned before called the pineal "pinecone" gland. The pineal has no apparent use, but it is thought to have been used by our brains at one time. After all, the gland didn't just grow there for no reason, so let's travel back to the Tower of Babel.

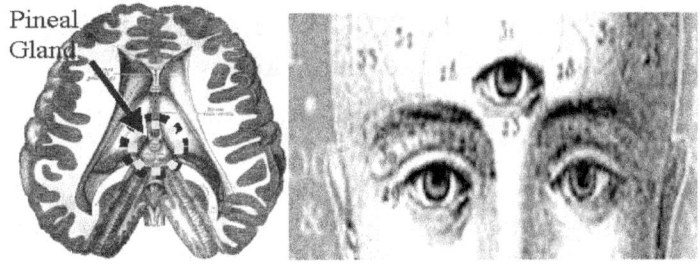

Tower of Babel and Pineal

The Tower of Babel is a huge structure that King Nimrod had built about 6 thousand years ago. A couple of things that may surprise you is that the Tower apparently was built in the country of Lebanon at a place called Baalbek rather than in Iraq as you have been told and the tower was associated with a huge world war when 1/3 of all the people on earth were killed. No one really knows what happened, but what can be derived from substantial evidence is that, all of a sudden, our brains lost most of their capability according to many ancient texts and how this brain loss was probably from some DNA modifying bacteria or something similar. I could also bring out the unusual fact that our current brain size is smaller than our earlier cousins, Neanderthal. While that fact is well know, what is not recognized is that this reduction in brain size shows that our brains began atrophying from disuse about 6 thousand years ago. I could bring out the fact that the entire world was plunged into some type of Stone Age re-insurgence 5 to 6 thousand years ago and people seemed to become dumb as stumps for a while. The Maya, Egyptians and Indians all began new calendars The Biblical book of Jasher simply tells us that 1/3 of the people died, 1/3 of the people became like apes and 1/3 of the people were dispersed to places around the world because they could only talk to there close relatives. We can imagine that before this brain reducing started, we could do many things with our bigger brain we cannot do today. We can imagine that the pineal gland, prior to whatever happened 6 thousand years ago also was larger and might have been used by our ancestors. I could also bring up many other things that would make you wonder if the pineal gland used to allow us to do many things in the past, but I won't. Instead, let me tell you what this tiny, pea-shaped gland does.

- **Pineal Glands** in many non-mammalian vertebrates have a strong resemblance to the photoreceptor cells of the eye. Some evolutionary biologists believe that the pineal cells share a common were the ancestor to retina cells in the eye.
- **In some animals,** exposure to light of this gland can change the animal's biorhythm.
- **Some early vertebrate fossil skulls** have a pineal opening so that it probably had some vision characteristic.
- **The lamprey and the tuatara** both have this same type of pineal opening and this thing is photosensitive. The structures appear to include cornea, lens and retina,

- **The pineal gland is weird** in that it has profuse blood flow, second only to the kidney, so we can be sure that it once was of great importance. While doctors are perplexed at why this insignificant gland would need so much blood, it is obvious that whatever happened 6 thousand years ago made the extra blood flow unnecessary.
- **Potentially, this organ aided** our capability to vibration our conspicuousness to higher levels. All these brain vibration levels are not exactly the same as consciousness vibration, because these are physical elements that stand in the particle dimensional dynamo. However, there seems to be a correlation to the "Expected effect" to consciousness and the brain vibrations so we should not ignore the characteristics or the oddness of the pineal.
- **The brain of a 90 million year old bird** was found with a large parietal eye and pineal gland so it's been used for some time now to provide additional insight beyond normal seeing.
- **Production of melatonin** by the pineal gland is stimulated by darkness and inhibited by light. This melatonin stuff affects sex drive.

I know this is a little out of the way from Life and death, but we are talking about vibrational oneness of our consciousness and there are some elements that must be understood. We have a tiny organ that used to be huge and it used to be an aid in seeing, regulating moods and sex drive, but our bodies are still trying to supply it with enough blood to run a huge organ. Today, the tiny little thing seems to have been abandoned by our bodies, but maybe we just can't see what it can do without vibrating a little. Vibrating allows us to understand our world.

Thomas Maslow
Anyway, this pineal gland/third eye was supposed to have given us the ability to understand the world around us. If we increase our vibrational level by unison with our environment "some call it meditation" or by other exotic means, we can sometimes get in tune with the world around us and here is the odd part. We can

even affect it. Another way of saying this is that the 3rd eye thing is that "Self–Actualization" that Thomas Maslow talked about.

Positive Thinking

Somehow getting our vibrational levels in tune with the vibrational patterns of the elements around us allows us to be more intuitive. As the vibrations increase, less and less of the conscious experiences relies on the carnal experience. Less and less of our "Life" is established in what we consider reality. Let's think of this whole consciousness a little deeper. Let me start over with a question.

Can your consciousness REALLY leave your body? - ABSOLUTELY!!!!!

I'm sure your first thought is that it can't and even after I discussed nears death experience and other similar accounts of people leaving this conscious world you still just can't get it in your head. Don't be so ready to close your mind to things that seem to be going on around us. It is becoming more and more apparent each year that astral projection, near death experiences, prophets seeing beyond this reality, and even reincarnations have been and are elements of the same characterization of the consciousness dimension. I know you think you are using your consciousness right now, but there is more to it that you would like to believe. Others simply shake their heads and believe that it might be sacrilegious to even suggest such a thing, well! It is not!!! There is a reason so very many people have witnessed similar things. It is not because Satan enters one's body and forces the images into a consciousness. It is because that is how God made us. Let's take another look at the various well known accounts of vibrating out of one's consciousness.

Near Death Consciousness

It is believed that over 10 million Americans have had Near Death Experiences and lived to tell about it and most concur with the general light tunnel and all of that.

Check out [http://www.nderf.org] and read the accounts. The picture is similar and seems to go along with a 10 dimensional vibrate to counter entropy universe.

- **Many say they feel a** "Whoosh" as they go through a tunnel much like the Bosh artwork depicted on the cover. The Whoosh is reported by deep meditaters as almost a buzz as the vibration level of our being brings us to the brink of separation with our bodies. This is consistent with the 10-dimensional universe and the life vibrational pattern.
- **Many see an overwhelming brightness-** Of course, no one actually sees light. They only see vibration and understand that there is a brightness. As one opens up their awareness of the things beyond the Carnal existence, what we interpret should allows seem bright.
- **People say they FEEL the Light-** The warmth is oneness with the spiritual characteristic of the universe. I know a lot of this seems like some shaman waving his raccoon claw over your head, but the idea here is that electro-magnetic interactions and ethereal interactions have a duality. Feeling warmth and light are when the consciousness goes towards more "soul-like characteristics just like the magnetism and gravity components. Having life vibrations get slower toward entropy places us back to the low basics of those lower chakra things. When we go there, things get dark and colder. As one

vibrates to the love levels, companionship gains warmth and brightness. This goes beyond that characteristic.
- **People say they feel an intense love and feel totally at peace.** There is no doubt that love requires our consciousness to be removed from the lower vibrational levels so during these experiences there must be this loving feeling.

I know some people have had the opposite experiences, but that makes sense as well as vibrational levels can be forced slower to become more carnal if one really tries.

This Is Not the End
Hopefully you are beginning to see that the going to the light and feeling the warmth and all the rest may not be the end of carnal living. Nor is it a one way tunnel. All the warmth stuff, friendliness, and other feelings seem to all be their, but death is a very strange component of life. Let me finish the life part and I will get right on death in a more open minded way. While we are alive, let's also look at how easily the consciousness can leave the body [Alive or Dead].

Out Of Body Consciousness

It is estimated that about 1/4 to 1/3 of the population depending on which study you look at have experienced SOME type of out of body thing. Sometimes it is just for an instant, a feeling that you already did something or know something that will happen, but many of us lose linkage with the COMMON consciousness we depend on. Here is what people say again. Rather than simply saying all the witnesses were lying, understand that our "reality" is esoteric. Jesus told his followers to make wine out of water, heal the sick with their hands, walk on water and even bring some people back to the state we call life.

- **Not Dreamlike**-They insist that the time they are away is not dreamlike, but instead it is close to reality. Don't for a minute believe that the spiritual world is not a reality. It is simply a different reality and for most of us it is very difficult to even glimpse this wonderful part of our dual universe.

- **Feel Powerful**-They usually sense power and freedom. Like I keep saying, the way to gain these levels of insight is to increase one's vibrational level way beyond the entropy element of destruction and separation from God and the spirit world. That increase in vibration is what is ALWAYS associated with feelings of freedom from the rigorous nature of our "Collective consciousness.

- **Aware of the surroundings**-These people recognize and describe objects seen in these states with great accuracy. Others, including many who initially were very skeptical, have verified this strange fact. Don't let people try to tell you that heaven or other universes are removed from the carnal one. They are co resident in space but separated by conscious characteristics with time so there is no doubt that objects could be sensed as objects.

Projection

One type of out-of-body experience is call astral projection. In this method of increasing ones conscious vibrations, people indicate that they sometime prepare several days and focus on a place that they wish to project to and they have various techniques to place themselves in a hypnotic state. Sometimes a simple word or "Mantra" is used to set up a situation. Once the initial conscious level is elevated, they feel like they are flying and they must fight not to fall into a deeper sleep level. And they feel at peace with the universe.

Additionally, we find that sometimes communication with others can be established in this state. It is not clear if the individuals are always real or self generated, but there is a substantial amount of information that suggests that astral projectors can communicate with other people and sometimes these people are not alive.

A Common Thread

Hopefully, you are seeing that in almost all cases, people begin these experiences by blocking out the world including all feeling. Those who are forced in that condition by some tragedy don't seem to have any difference in this effect. They leave their bodies, get comfort or wisdom, sort of talk to comforting people, get a heightened sense of reality, can float, and when they get focused back on the "real world" they are plummeted back into it. Many times these people are changed forever. I think this is something close to the crown chakra level that ancients attest to. The trip nearing the Crown chakra has changed them forever and I'll tell you why. Their consciousness has been vibrationally enhanced. It vibrates closer to the level needed to do this transfer thing, but in the mean time, they become more aware of the feelings of others and become more self-actualized. Whether the "people they interact with are the cause of the vibrational enhancement or some other mechanism is at work, I do not know, but the entire life force of the person is enhanced. Many like it so much they go off and do it again if they can.

These astral projections and out of body episodes bring people closer to the vibrational level associated with the crown level, but one more thing is required to go all the way to this heaven place. Before I get to that, I think I had better figure out things about Carbon 12 "the building block of life" before we get to the death section.

Carbon-12 Impossibility

Today scientists can't even figure out how carbon-12 was made, much less many of the larger more complex atoms. Carbon-12 is made from Helium-4, and Beryilium-8. That seems straight forward except beryllium-8 is extremely unstable and lasts only 0.00000000000000001 or 10^{-17} seconds before it turns back into helium-4. It is not enough time to produce the combination. Even today, no explanation of how carbon-12 is made has had any real meat.

Carbon-12 seems to not be possible, but there Carbon-12 is building people.

It is as if people were here and existence went backwards in time from our existence. I know this still sounds odd to you because you are used to time going in one direction and that no one or nothing can go backward in time---especially not a universe. It would also mess up the next section of the book as death would be the moment of inception which I certainly am not getting into.

I know some of you are skeptical about your being able to shape reality with your thoughts and you don't see how that ability can affect your death but hopefully, after you read this book, it will be clearer to you. If you were skeptical about life, I guarantee you will have a somewhat difficult time with the subject or death. Just to be clear let be say that I'm going to discuss how death fits into a vibrational world where even our conscious thought can change the very framework of our existence. Even after death, there is more to learn. For the most part, I will examine both external characterizations and religious discussions so we can compare details as we go.

Ascent of the Blessed by Hieronymus Bosch

This 15th century depiction of conscious "spirits" of people passing from this world to another through a huge lighted tunnel was the depiction 600 years ago and it is still the belief of many. Death is a one way trip to happiness or doom. Most sense this lighted tunnel thing during "near death experience" which further intensifies the belief, but it seems there is much more to it.

Death in Religion

As I mentioned, this book is about the science of life/death, and reality, but is also uses a religious base to comingle with the science. If science and religion did not agree, there must be something wrong with one or the other. Let's start by understanding what the various religions of the world believe. This section will not get into the anthropic nature of the conscious, but I think you will start to see a pattern within the various religions of the world that we can use as a base of sorts. I'm certainly not going to do them all, but I think a spattering will help us approach this subject better and I am going to concentrate on Jewish and Christian beliefs quite a bit, because we can get some very good basics in our attempt to understand what death is all about. If you don't want to be bothered with religion, please stick it out as there are some important concepts in some of these ancient texts that will help us understand the Physics of death as an energy transfer rather than some possibility that the energy of a soul somehow is terminated. It simply cannot happen..

Taoist Belief

In Taoism, there is not exactly an afterlife. While that sounds final, they believe that we are eternal in Tao. The afterlife is, sort of within life itself. People are of the Tao when they are living and when they die, they are the Tao. I know it sounds like there is an afterlife, but, I talking about the Taoists here and they don't like to think that way.

Hindu Belief

Hindu believe in many Gods controlling different segments of life. Hindu have this doctrine of "anatta", which is the notion that individuals don't really possess eternal souls. Instead of eternal souls, we consist of a bundle of habits, memories, sensations, and desires. Together these things make us think we are in a stable, lasting self. This false self hangs together as a unit and reincarnates in body after body. The goal of this glob is to obtain release by abandoning the false sense of self. They also have this belief that reading of the six Shastras and their mastery will bring salvation. In other words, man controls his destiny which is opposite to the Christian belief that man cannot, ultimately, survive after death without God's Holy Spirit and forgiveness. The "ultimately" part is something we are going to address in this book to allow for a much clearer image of death.

Buddhist Belief

Buddhism does not believe in the need for God, in that sense Buddhist doctrine is absolutely atheistic. However, Buddhist have a similarity to both Tao and Hindu in that they have doctrines of reincarnation, karma, the notion that the ultimate goal of the life is to escape this thing they called the cycle of death and rebirth. They believe we are bound to the death/rebirth process by desire or craving anything in the world. Depending on your special nature, you may not come back as a person. Instead, one may be a rat of cow. I don't know if anyone comes back as a roach, but be careful just the same. Free yourself of desire, and wham! You are free to be in something they call Nirvana or "liberation"--- and you don't have to worry about being a bug. Some of the more strict believers practice something called "self mummification" which requires the person to turn his flesh to stone with various poisons and tree bark over a period of many years. If all goes well, when the crazy person dies, he is mummified and is given the name of Buddha. Because he was able to completely separate himself from the CARNAL life, he was relieved of the life/death cycle.

Moslem

The Moslem religion is strange. Maybe we will get some insight from the Quran [Chapter 4 verse 34]. *"Men have authority over women because Allah has made the one superior to the other, and because they spend their wealth to maintain them. Good women are obedient. They guard their unseen parts because Allah has guarded them. As for those from whom you fear disobedience, admonish them and send them to beds apart and beat them."* Never mind! If a religion can't even understand that people are people, we won't find out much truth. I tried that beating my wife once, but she assured me that I was not a Moslem and bonked me on the head.

Kemetians Belief

The ancient Egyptians or Kemetians told us some about their belief in death, but let me share with you their belief about the beginning first because this is older than the Old Testament and I want you to see similarities in concept as we start putting together the similarities to support a more reasonable concept of death. This comes from very ancient Egyptian texts.

"On the second day was born the great Elder Horus. On the third day was born dark SET [or Satan, also known as Gadrael in Jewish texts].

On the fourth day was born Isis; on the fifth day, Nephthys, thus is the birth of the Great Shining Ones [or Archangels], the Company of the Gods of Annu." [While the Kemetians called them Annu, Sumerians called them Anunnaki, and the Jews called them Anak or Anakim. The Anak, according to Jewish texts, were the "Losers of a War in Heaven" who became human and ruled what was left of the Earth after being totally destroyed because of the conflict.]

"Many other Gods/angels were also created by God; and He filled the sky above the Earth. Last of all was created Man and the other beings of the Earth. The mighty Khnem'u fashioned them

upon His potter's wheel, and Re breathed into them the breath of life." [Just like the Jewish version, the Heaven War occurred well before a new version of man was created from dirt on the 6th day.]

"Further, He made a land for them to dwell within, and named the kingdom Kemet. He populated the world with all forms of animal, bird, fish, and plant; and gave them also the breath of life." [Just like the second chapter of Genesis, animals were re-created after the Heaven War.]
"The forces of darkness were not conquered forever at the beginning of time; instead they surrounded the Earth as serpents poised to attack God. The war between darkness and light sustains the world; and when it comes to a final end, so too will the world. [Just like the Biblical, these Anak characters are still around and still upset about losing that war during the very ancient times. They and their followers will be sort of "reinstated" and fight another huge war which will cause the Earth to be destroyed.]

The reason I brought out this section of their belief will make more sense later as we look at the punishments of the Anak that concern death. If we know how other entities died, it will give us a more reasonable picture about how we will die.

In other texts, the Kemetians believed that when they "died", *"their soul came to the place it knew, and it would not overstep the ways of the past."* I'm not sure exactly what that means, but it sounds like your soul learns each time you die. They also indicated, *"A person survives after death, and past deeds are laid before the person in a heap. If your heap is not big enough, you go to a bad place"*. They believed that existence in the "afterlife" is eternal, and one who "measures up" will be like a god, striding forward like the "Lords of Eternity".

Inca Belief

Across the Ocean, the Inca worshipped the dead, ancestors and founding culture heroes. Their king, for instance, had attained divinity by heritage rather than being good. While that was somewhat different, like the Kemetians, the Inca wanted to regain their bodies in the "afterlife" so they mummified dead rulers, children that had been sacrificed, and a group known as the "Cloud people". At some point in time, the "wanderers" could regain their bodies. Just like the Christian belief of having the dead bodies come out of the ground during the second coming of Jesus "God incarnate", the Inca believed in this transition into a new life in the future.

Aztec Belief

The Aztec spiritual belief consisted of preparation, fasting, purification, and offerings. After this came blood sacrifice. This wasn't like the Jewish slaughter of a Goat. These guys used captured warriors, woman, and children. This is the important part. For them there could be no new life without death. The Aztecs also had their own version of Hades and even the grim reaper called Mictlantecuhtli, This guy looked into the deepest part of your soul and you didn't want to make him angry because he was the ruler and creator of the Aztec Underworld.

Greek Belief

The ancient Greeks believed that at the moment of death, a person obtained a higher level of consciousness. These guys believed that all souls passed on to the underworld realm known as Hades. Not just those who had been bad during life, I'm talking about everyone. If you were bad, you went to Hades basement. Tartaros was a realm below Hades where disobedient people and gods were sent for punishment. While Hades was sort of a purgatory, Elysium was a beautiful "heaven-like place" that was inhabited by those Zeus favored. The Greeks had great influence on the Jews as we will see later.

Sikhism and Eckankar

Sikhism is one of the newer religions and started about 500 years ago in India to "remove issues with Buddhist and Hindu teachings". Today 20 million people worldwide have this as their belief, so it is the worlds 5th largest religion. With that many people believing it, we might get good information about death.

Sikhism preaches a message of Devotion and remembrance of God at all times, truthful living, equality of mankind and it denounces superstitions and blind rituals. Eckankar is a newer, more Westernized, version of this Sikh religion, so we might get a clearer perspective. Eckankar focuses on spiritual exercises that enable people to experience something called "the Light and Sound of God." This experience is the primary goal of the teaching, just like the Christian belief is to obtain the Holy Spirit that is characterized as the Light of God. This light is the spiritual inner path to understanding of self as a soul and it assures a development of higher awareness of God. I know that sounds just like Christianity, but there are major differences coming up.

Once one attains the "light", he becomes an *Eckankar* or "Co-Worker with God". Believers in Eckankar, like Hindu, Jews, Buddhist, and many others don't believe in the incarnation of God as Jesus. He was merely one of the people who had gotten this light stuff. Some simply say that the Eckankar [those who attain the fullness of understanding] are COWORKERS or God or a version of God. To show how important becoming an Eckankar is, the word ECK means "the Holy Spirit of God".

The second major teaching is that death is transitional. One passes from life to life trying to attain this "co-worker" status by learning not only while one is alive, but also between lives by listening to "teachers" who have attained that "co-worker" status. There are countless cycles of births and deaths. One only breaks this cycle when he achieves "mukhti" [merger with God]. Similar to Buddhist Chakra levels, something called Karma regulates the reincarnation and transmigration of the soul so in Sikhism and Eckankar, they link Karma with the doctrine of Grace used by the Christians. While this sounds somewhat sacrilegious, there is a closeness of religious thought. To show how little difference there really is, they teach that mortals obtain a human body because of good deeds but they can only reach the gate of salvation with God's kind grace. With that, let me switch over to Jewish religion and expand what people think a little more. By the way, I know there is quite a bit about religion in this book, but religious tell us more about death than the normal history books and scientific research, so we simply cannot ignore these inputs if we are trying to understand something like DEATH.

Jewish Death

I put this section together to establish some position on death in terms many of us already understand. While some of this next set of verses is specific to the various Jewish religions, it seems reasonable to put them together so that we can see transition of thought and compare them to the other ancient teachings from around the world. While reading these things remember that the energy of life cannot simply be extinguished no matter how difficult it is to understand what a transition from life to something we call death can be. Our soul makes up a piece of our reality.

Ecclesiastes

First, let's look at the book of Ecclesiastes. Chapter 1 and 3 tells us the following. *"What has been will be again, what has been done will be done again; there is nothing new under the sun ... Whatever is has already been, and what will be has been before; and God will call the past to account."* Certainly, reincarnation is discussed here, and in the Jewish belief, God would remember all of your past lives as part of the full accounting to see if you went to a good place.

Sheol versus Hell

Hopefully you are seeing a trend in other religions and the one you thought you already knew about. Rather than believing in a Hell, early Jews believed in a place called "Sheol or purgatory" an area beneath the Earth where people went after death. It was neither good nor bad, just a place to go between reincarnations.

Later we will look in the book of Luke to get a good picture of the Sheol place.

Sadducees and Zoroastrians

Over time, the Zoroastrian beliefs of "heaven", "hell" and the dual nature of mankind and the final rule of God on earth started to push into the Jewish religion. The Zoroastrians were descendents of the Chaldeans of Persia who had incorporated some of the very ancient Jewish religious components into their own religious views after holding them captive for many years and this melded composition was now invading the land. This slow conversion of the population of Jews made some people uncomfortable. A sect of purists, called the Sadducees, which was made up of over 90% of the population, rose up to reject all Persian concepts including resurrection, angels, and spirits. According to the Gospel of Matthew [chapter 22 verse 23], the Sadducees generally believed that after death, there was this Heaven/Hell thing, but how you made it to one or the other was a little sketchy. Sadducees weren't much better in their understanding of death.

Pharisees

A group of the Jews started reading the "Holy Books" again and determined the Sadducees might not be reading everything correctly. This group became known as Pharisees. The Apostle Paul was one of these before he became Christianized. Unlike the Sadducees, these Jews did believe in reincarnation. They believed the souls of evil men were punished after death. The souls of good men were *"removed into other bodies" and they "had power to revive and live again."* Still the Carnal life was the end goal of these Jews-- Not Heaven and not going to be with God. Possibly Sheol was involved somewhere, but mostly they were afraid of death.

Essenes and Pythagoras

The Greek and Pharisee religion came together and produced the Essenes. They believed in reincarnation and like the Buddhists, non-violence to all living creatures. You just never know who you might be eating. Many of the Christian dogmatic customs also took their root from the Essene practices, before our Bible was put together. Both the practice of baptism and of the sharing of the wine are recorded as part of their practices long before Jesus and John the Baptist popularized them. This strange group was, sort of, followers of Pythagoras, the Greek philosopher who taught reincarnation. The Essene believed that the soul was both immortal and pre-existent. Oh Boy, pre-existent is much more than reincarnated. Our souls were from the beginning of time. Another thing odd about the Essene is that they had a book about the end of time. Pieces of it have been found and it speaks about what happens when you die. This book sounds identical to the "Revelation" book in the Bible and it was written well before the time of John who wrote the later one.

"The Essene Book of Revelation"

It is commonly called "The Essene Book of Revelation". Let's see what it says about dying. *"And then I looked, and behold, a door was opened in heaven: And a voice which sounded from all sides, like a trumpet, spoke to me: "Come up here, and I will show you the things which must be hereafter."* [Our Bible and the ancient Essene texts are filled with these transmigration events. In many texts we find, the spirit was able to see what the "heaven universe" really was like.]

And immediately I was there, in spirit, at the threshold of the open door and I entered through the open door into a sea of blazing light. And in the midst of the blinding ocean of radiance was a throne: [I know this sounds just like the descriptions from "Near-death Experiences", so we might believe that people are not simply saying they saw this huge light because others said they saw it.]

And on the throne sat one whose face was hidden. And there was a rainbow around about the throne, which looked like emerald. And round about the throne were thirteen seats: And upon the seats, I saw thirteen elders sitting, clothed in white raiment; And there faces were hidden by swirling clouds of light. And seven lamps of fire burned before the throne, The fire of the Earthly Mother. And seven stars of heaven shone before the throne, the fire of the Heavenly Father. [The rainbow around the head, the swirling clouds of light and the other things seem to indicate that light is self generated rather than photonically generated as we teach in the carnal world. The whole concept of the Watchers [or angels] glowing when people saw them and how Moses' face began to shine and how Noah's face shined when he was born all seem to tell us the same "light" is in this universe, but more easily rendered after carnal death.]

And before the throne there was a sea of glass like crystal: And reflected within it were all the mountains and valleys of the Earth, and all the creatures abiding therein. And the thirteen elders bowed down before the splendor of him who sat upon the throne, whose face was hidden, And rivers of light streamed from their hands, one to the other, And they cried, "Holy, Holy, Holy, Lord God Almighty, [This crystal that could see into our carnal universe from the Heaven universe is very interesting. Just how one could see across the two universes is still a mystery, but it does give us new understanding about the connection of death, or, at least, the integration of a universe for the dead and one for the carnally living.]

I Samuel Awakening the Dead

To make this being conscious when dead thing even more apparent, let's look at I Samuel 28:15-19. This witch awoke up the spirit of Samuel to talk to King Saul. Samuel wasn't happy about it, but he does listen to Saul. *"Samuel said to Saul, "Why have you disturbed me by bringing me up?" "I am in great distress," Saul said. "The Philistines are fighting against me, and God has turned away from me. He no longer answers me, either by prophets or by dreams. So I have called on you to tell me what to do." Samuel said, "Why do you consult me, now that the LORD has turned away from you and become your enemy? The LORD* *has done what he predicted through me. The LORD has torn the kingdom out of your hands and given it to one of your neighbors- to David. Because you did not obey the LORD or carry out his fierce wrath against the Amalekites, the LORD has done this to you today. The LORD will hand over both Israel and you to the Philistines, and <u>tomorrow you and your sons will be with me</u>. The LORD will also hand over the army of Israel to the Philistines."* Here we find that good people simply sleep when they die and they don't go to heaven immediately. Therefore, they can be easily awakened.

Jeremiah

Jeremiah 1:4 says even more about existence before birth. *"Then the word of the LORD came unto me, saying, before I formed thee in the belly I knew thee; and before thou camest forth out of the womb I sanctified thee, and I ordained thee a prophet unto the nations."* This is talking about life before the womb or what we might call cognizant death between lives if we are reading a book on death.

Malachi Reincarnation

The Old Testament prophets, such as Malachi 4:5 indicated that Elijah would return before the Messiah would come. *"Behold I will send you Elijah the prophet, before the coming of the great and dreadful day of the Lord."* [Reincarnation]

Job Reincarnation

Job 1:20-21 fills in more detail. *"Then Job arose and tore his robe and shaved his head and he fell to the ground and worshipped. And he said, "Naked I came from my mother's womb, and <u>naked I shall return there</u>. The Lord gave and the Lord has taken away. Blessed be the name of the Lord."* Job says here that he will return to be born again in a womb indicating that he expects to reincarnate.

More Job Reincarnation

Job 19:25 speaks even plainer as he indicates that he would get new flesh and new eyes "BEFORE" he saw God again. *"I know* *that my Redeemer lives, and that in the end he will stand upon the earth. And after my skin has been destroyed, <u>yet in my flesh I will see God</u>; <u>I myself will see him with my own eyes-I, and not another.</u>"* [Reincarnation]

Ecclesiastes Reincarnation

Ecclesiastes 1:8 tells us even more. *"What has been will be again, what has been done will be done again; <u>there is nothing new under the sun.</u>"* If there are no "new babies", the entities must have been here from the beginning of time.

More From The Essene

If living before one is alive is a strong belief so must be reincarnation that would allow for it. Dead Sea Scrolls helped us understand more about the ideas about death by the Essene Jews. One particular Dead Sea Scroll Called "Melchizedek Text" mentions reincarnation as the great priest *"Melchizedek is reincarnated in the last days to destroy Belial (Satan) and lead the children of God to eternal forgiveness"*. The current New Testament books indicate that the one who destroys Satan is Jesus during his final return, so one may believe that Jesus had been reincarnated from the beginning and he was originally this Melchizedek guy.

Gnostics and Plato

Instead of going with Pythagoras, one group began combining Jewish beliefs with those of Plato. These were the believers of "transmigration." Unlike the preexistent soul, this group believed that one could allow your "Soul" to migrate out of your body even before you were dead. OK! I just went through and showed that our Bible and the Essene both describe this transmigration thing, but this is one of the reported differences. I suppose you could say that the Gnostics believed that you could transmigrate often if you wanted to and the Christians and Essene indicated that only special circumstance allowed for it.

Modern Jewish Belief

With all of these mixed up thoughts, modern Jewish ideas include the concept that people could live again without knowing exactly the manners by which this could happen. They firmly believe that death was not the end of human existence. However, because Judaism is primarily focused on life here and now rather than on the afterlife, Judaism does not have much dogma about the afterlife, and leaves a great deal of room for personal opinion. Therefore, some believe the "resurrection" refers to a time when souls of the righteous dead go to a place similar to the Christian heaven. Others believe the "resurrection" refers to the reincarnation of a soul through many lifetimes.

One of the Jewish "rule books" called Zohar states the following: *"All souls are subject to revolutions." [Reincarnations] "Men do not know the way they have been judged in all time."* (Zohar II, 199b) Therefore, Jews would be judged for bad things and have no memory of the things they were judged for in the Afterlife. This apparent erasing of memory is a very important concept so remember it just a little.

Another book "Kether Malkuth" tells us that "*If the soul is pure, then it shall obtain favor. If it has been defiled, then it shall wander for a time in pain and despair until the days of her purification.*" You might wonder, "How can the soul be defiled before birth?" Jewish Rabbis help and say this verse means that the defiled soul wanders down from paradise through many births until the soul regained its purity. My feeling is that Jews really don't know what to believe.

New Testament Death

I picked out a number of the New Testament passages, for two reasons. The first is that many don't know what the Bible has to say about death and the second one is that the sayings of God incarnate, Jesus, give a high level of authority on what we might gain. This is not a religious book per se, but there will have to be a flavor of religion in that matters of death are not easily described without it. That being said, the early Christians were all mixed up. They had a strong belief in reincarnation and they weren't sure about how many Gods there were. This is strong belief in reincarnation messed up their belief system. To show how very deeply the early Christians believed in reincarnation, Jesus was believed to have been the reincarnated Melchizedek by some and Isaiah by others. It seemed that almost none of the early Christians believed that Jesus was simply born by Mary who was inseminated by God. Some believed Jesus was simply a wandering spirit/soul of the earlier great men. Because Jesus had gained a new level of enlightenment, they believed he could be presented as the "Son of God." This is much like the Eckankar belief. Later, Christians started reading the Bible and realized that Jesus was something more special than the wandering spirit and his Holy Spirit was the key to another level of death or life after death or whatever we want to call it. The early Christians certainly were looking for other explanations because they didn't even have a Bible.

Matthew Reincarnation

Chapter 11 verses 13 and 14- "*For all of the prophets and the law have prophesized until John. And if you are willing to receive it, He [Jesus] is Elijah who was to come.*" People believed that Jesus was simply the reincarnated Elijah.

More Matthew Reincarnation

In Matthew 17, Jesus, again, identifies John the Baptist as the reincarnated Elijah. "*And the disciples asked him, saying, 'Why then do the scribes say that Elijah must come first?' But he answered them and said, 'Elijah indeed is to come and will restore all things. But I say to you that Elijah has come already, and they did not know him, but did to him whatever they wished. So also shall the Son of Man suffer at their hand.' Then the disciples understood that he had spoken of John the Baptist.*"

Still More Matthew Reincarnation

After the death of John the Baptist, whom Jesus identified as Elijah, Elijah appears again along with Moses at the Mount of Transfiguration. (Matthew 17:1-13*)* "*After six days Jesus took with him Peter, James and John the brother of James, and led them up a high mountain by themselves. There he was transfigured before them. His <u>face shone like the sun</u>, and his clothes became as white as the light. Just then there appeared before them Moses and Elijah, talking with Jesus. Peter said to Jesus, "Lord, it is good for us to be here. If you wish, I will put up three shelters-- one for you, one for Moses and one for Elijah." While he was still speaking, a bright cloud enveloped them, and a voice from the cloud said, "This is my Son, whom I love; with him I am well pleased. Listen to him!" When the disciples heard this, they fell facedown to the ground, terrified. But Jesus came and touched them. "Get up," he said. "Don't be afraid." When they looked up, they saw no one except Jesus.* So, these dead people Elijah and Moses come back to life, talking and completely visible, but then they vanish. These guys had been dead for hundreds of years. They had not been to heaven because Jesus told everyone that he was going to make heaven livable so you ask, "Where were they?" Well the next section tells us a little bit concerning Elijah and where he has been.

As they were coming down the mountain, Jesus instructed them, "Don't tell anyone what you have seen, until the <u>Son of Man has been raised from the dead</u>." The disciples asked him, "Why then do the teachers of the law say that Elijah must come first?" Jesus replied, "To be sure, Elijah comes and will restore all things. But I tell you, Elijah has already come, and they did not recognize him, but have done to him everything they wished. In the same way the Son of Man is going to suffer at their hands." Then the disciples understood that he was talking to them about John the Baptist." I know this was a long one, but it really brings out a lot about death, or what we call death. These guys Elijah and Moses had been great leaders who came back to life not only as a reincarnated, new person or persons, but also as themselves in sort of a "halfway" life on this mountain.

Ephesians Prenatal Existence

Ephesians 1:4 tells us the same thing. *"He chose us in him before the foundation of the world, that we should be holy and without blemish in his sight and love."* While this is specifically talking about Jesus here, other verses show that there was a belief in prenatal existence by all people.

Romans Prenatal Existence

Malachi 1:2-3 and Romans 9:11-13 both say more. *"God loved Jacob, but hated Esau even before they were born.* [Prenatal existence]

Mark and Luke Reincarnation

"But I tell you, Elijah has come." (Mark 9:13) and *"... the spirit and power of Elijah." (Luke 1:17)* --I know some try to tell you that the New Testament belief is that when you die, you go into the ground and wait for thousands of years only to be yanked up when Jesus comes again and meet him in the sky. While it does state that in a number of places [Daniel, Thessalonians, Revelation and other places] The idea that everyone is just sleeping all of that time does not sound like the belief presented

by Jesus. We can be pretty sure John carried Elijah's living spirit because Jesus, Mark, Luke and others said it, but according to his own words, he could not remember the things Elijah remembered

More Reincarnation

In all three of the synoptic gospels, Jesus promised that anyone leaving their homes, wives, mothers, fathers, children, or farms to follow him would personally receive hundreds more such homes, families, and so on in the future. Jesus said: *"No one who has left home or brothers or sisters or mother or father or wife or children or land for me and the gospel will fail to receive a hundred times as much <u>in this present age</u> - homes, brothers, sisters, mothers, children and fields ... and in the age to come, eternal life."* (Mark 10:29-30) I know you are wondering how someone can get all new brothers, sisters, mothers, and children without getting a reincarnated body during the "present Age". I know the book of Job sounds like everyone he loved was killed and he lost everything, then he got more than he had before without dying, but that is not the NORMAL way people got back everything they lost. Most had to die first. They didn't live another 160 years after everyone in their family had been killed like Job.

John Prenatal Existence

The epistle of John and John's Revelation both provide us with insight into dying. Here are a few excerpts. As we continue in this investigation, let's look at details of a blind man being healed in John 9:1. In this verse, we find out more about death or pre-life in something we could call prenatal existence. *"And as he was passing by, he saw a man blind from birth. And his disciples asked him, 'Rabbi, who has sinned, this man or his parents, that he should be born blind?" Jesus answered, 'Neither has this man sinned, nor his parents, but the works of God were to be made manifest in him.'* We are confronted with a revealing question. When could he have made such transgressions as to make him blind at birth? The only conceivable answer was that some prenatal problem had caused it. This was the normal belief. The question explicitly presupposes prenatal existence. It will also be noted that Jesus does not indicate that their ideas were incorrect, for most occurrences.

John 1:21 thru 27 Reincarnation

It is talking about John the Baptist being reincarnated. *They asked him, "Then who are you? Are you Elijah?" He said, "I am not." "Are you the Prophet?" He answered, "No." Finally, they said, "Who are you? Give us an answer to take back to those who sent us. What do you say about yourself?" John replied in the words of Isaiah the prophet, "I am the voice of one calling in the desert, 'Make straight the way for the Lord.'" Now some Pharisees who had been sent questioned him, "Why then do you baptize if you are not the Christ, nor Elijah, nor the Prophet?" "I baptize with water," John replied, "but among you stands one you do not know. He is the one who comes after me, the thongs of whose sandals I am not worthy to untie."*

Like the Essene Revelation, I brought out earlier; John's Revelation from the Bible gives us just a little more insight to build on.

Revelation 20: 1-6

Revelation starts telling us when the rollercoaster of life after life will change. It starts scary. The world had just come through a horrible time when the world had gotten more evil than it is today. *"I saw an angel coming down out of heaven, having the key to the Abyss and holding in his hand a great chain. He seized Satan [also called god of this world], and bound him for a thousand years. He threw him into the Abyss, and locked and sealed it over him, to keep him from deceiving the nations anymore until the thousand years were ended.---- And I saw the souls of those who had been beheaded because of their testimony for Jesus and because of the word of God. They had not worshiped the beast or his image and had not received his mark on their foreheads or their hands. They came to life and reigned with Christ a thousand years. This is the first resurrection.* [Notice that these "Souls" were seen on the earth, not in Heaven. They were simply walking around and came back to life. Later, everyone else comes back to life, but this is a good example.] *Blessed and holy are those who have part in the first resurrection. The second death has no power over them, but they will be priests of God and of Christ and will reign with him for a thousand years.*

Luke 16:22-31 Conscious death

If we really want to sense death, we simply have to look at Luke 16:22-31. This is the story of a rich man and a beggar named Lazarus that died about the same time. *"The time came when the beggar died and the Watchers carried him to Abraham's side. The rich man also died and was buried. In Sheol where he was in torment, he looked up and saw Abraham far away, with Lazarus by his side. So he called to him, 'Father Abraham, have pity on me and send Lazarus to dip the tip of his finger in water and cool my tongue, because I am in agony in this fire.' But Abraham replied, 'Son, remember that in your lifetime you received your good things, while Lazarus received bad things, but now he is comforted here and you are in agony. And besides all this, between us and you a great chasm has been fixed, so that those who want to go from here to you cannot, nor can anyone cross over from there to us.' He answered, 'Then I beg you, father, send Lazarus to my father's house, for I have five brothers. Let him warn them, so that they will not also come to this place of torment.'* Nasty, nasty! It sounds like if bad happens when you are alive, good will happen when you are not alive. We'll discuss that a little more but the main thing I wanted to bring up is that neither of these guys was in Heaven and both were conscious when they were dead.

Acts

Now let's look at Acts 2:38-39- *Peter replied, "Repent and be baptized, every one of you, in the name of Jesus Christ for the forgiveness of your sins. And you will receive the gift of the Holy Spirit. The promise is for you and your children and for all who are far off--for all whom the Lord our God will call."* This includes people alive and people far off [not alive]. Wow! This is talking about dead people being able to be revitalized with this Holy Spirit. This Holy Spirit thing is different than the other religions, but most of the rest is the same. <u>Reincarnation, Transmigration, Prenatal awareness</u> and all the rest was preached until the Catholics took control.

Catholic Rebellion

Something happened when Constantine I started building a new Christianity. The Catholic Church outlawed and put to death those that preached reincarnation and quit emphasizing the only difference in Christianity call the Holy Spirit. If you died before getting this marvelous thing, they came up with something called purgatory, but they used it to get money rather than using it to help people understand the basic characteristics of death [Reincarnation, transmigration, the Holy Spirit, Heaven/Hell, and prenatal existence]. It wasn't long before much of the Christian beliefs were tainted. The fifth ecumenical council in 553 AD stated the following: *If anyone asserts the fabulous pre-existence of souls, and shall assert the monstrous restoration which follows from it: let him be anathema.* This anathema word means "something dedicated, especially dedicated to evil" so it was a really bad word to use.

After Death Description

Hopefully, you are still reading the book. I had to establish the precedence of religions you understand so we can look at each of the elements with a level of authority. In the last book on Vibrational Matter, I indicated that I didn't really know what happens after death, but there is a peculiar thing we recognize from Biblical texts that sometime we ignore. At the beginning of this book, you may think I am a Godless, atheistic, egocentric, nilist that has no business telling anyone anything as it might disturb the natural order. Hopefully, you will change you mind as we go along as I am strongly religious and use the oneness of religion and science to help build a truer understanding of the mysteries of the Bible and science. For instance, from the last book of the Bible "Revelation of John" we find still another very important element of living when we are dead. ---

"When God returns to the Earth in glory, the dead in Christ will be raised".

Certainly one cannot come out of the earth unless one is in the earth so we can first see that some people or consciousnesses are sort of "sleeping" for thousands of years. All of a sudden, these people wake up and rise up into the air. This is only one of the mysteries we will try to understand. We will not simply try to talk away the details in an effort to conform to dogma of a Church, but wee will, instead look at the words---novel as it might be.

As I stated before, some try to twist the words around and indicate that only dead, moldy, bodies are raised up out of the dirt in the last days. Somehow, they think that there is some desire in our consciousnesses, which have gone to ANOTHER PLACE for thousands of years, to get into the exact dirt bodies we had eons before. I know that doesn't make sense, but MANY people use this example and the raising dirt bodies as an explanation so that they can be this backdrop belief that when people die they immediately god to Heaven or the bad place. Hopefully you are beginning to see that our great religious works don't say that at all.

God Does Good By Destruction

The second thing that I need you to accept from our Bible and many other sacred religious texts is the following. This is a major thing to understand for this book and for your own understanding of Life and this transitional thing we call death.

God causes everything to work together for the good of those who love God [Romans 8:28]

This "good" includes mass destructions, and telling the Jews to kill all the Amorite, Hittites, and Jebusite children, women, men and animals. I'll explain how a better DEATH definition makes this statement more palatable. To make this strange request by God even stranger, the Jewish people were to allow the Egyptians to live and work with them while many other nations were identified as needing to be completely annihilated. The Egyptians had enslaved, beaten, killed, and tortured them for hundreds of years while completely hating the one true God. The other groups were related to the Jewish people more closely and had similar lifestyles to the Jews. Could God have told them the opposite of what he wanted?

Judas Never Born

Another verse I brought up was a statement from Jesus. He indicated that ----

"It would have been better if Judas had never been born"

This was talking about one of the disciples who had been worshiping with Jesus for years and had finally betrayed Jesus, almost the same as Peter did by denying that he even knew Jesus during the trial of God incarnate or Jesus. Some may claim that Judas Iscariot **had** to do all that betraying to satisfy the scriptures. No matter what the secret reason may have been, we are told that Judas simply wanted Jesus to free the Jews and his backsliding was only because Jesus was not "apparently" helping the Jews quickly. He was not an evil man so the wording is curious and seems to be talking about the pain Judas would suffer for his indiscretion. He finally hated what he had done so much that he hanged himself. **[It would have been better for Judas to not have been born because of the massive guilt he had to bear.]**
At one point the Babylonians, under King Nebuchadnezzar, were said to be doing God's work as they slaughtered many in a power crazed effort to control the entire world. Even the greatest Apostle of God, Paul, was trying to single-handedly destroy all of the Jesus worshipers before God blinded him on a roadside. Why had God not said, "It would have been better if Nebuchadnezzar and Paul had never been born"? This will make more sense as we go along as well and when we add the physics to it, it all comes together so we don't have to ignore religion to understand science and vice versa.

No Divine Insight?

This detail is not coming from some flash of enlightenment that others claim to have concerning insights that no one else can fathom. I'm not saying God does not give divine wisdom if it is needed, I'm simply saying that most of the time, what people this is divine insight is nothing more than gas. I'm not trying to be coy or sacrilegious, but just look around. Hundreds of religious leaders study and seem to think they get heavenly guidance and contradict each other on every turn. God could "guide" only one of the 4,326 or so heartfelt opinions, presented on any subject.

The other "opinions" of thousands of truly devout believers in God MUST BE wishful thinking. These devote people who truly want to help everyone they talk to, ARE NOT GETTING INSIGHT from God. Remember, God is the controller of hurricanes, floods, volcanoes, destruction, sickness, and death. At the same time, he does everything for the common good. It sounds so strange when represented by most of the religious leaders of the day. We will examine how, why, when, and where God's control, love, and mercy MAKES sense. We not only can accept that there are miseries, but also that the miseries are TRULY for the GOOD. OK! I hope some of the information I will be trying to provide will have a hint of insight, but you will have to get the details ironed out on your own as there are some Preston-isms that simply can't be helped.

Saul and Samuel's Ghost

Before we start getting into the physical laws, theories, and evidence, let's continue by looking at another very important Biblical detail I brought up in the last chapter. When the Israelite King Saul needed insight into a war that was brewing, he had the dead Judge of the Israelites named Samuel ----

"was awakened from death to tell him what was to come in the future".

Samuel woke up in a spirit form and told Saul what was to come about in the near future and complained that he had been awakened.

From this we can imagine that, after death, there is some level of consciousness, <u>some portion of the future is known</u>, there is some connection between living and death dwelling places, and people typically like being in the Dead place rather than in the CARNAL universe.

Kill The Amorites Enigma

We will look at other details later, but one might wonder, "Why would there be a consciousness for thousands of years and why would killing entire civilizations be for the common good—especially considering the Amorites were part of the common good?" To refresh you memory, here is a snippit from Deuteronomy

> **"Do not leave alive anything that breathes. Completely destroy them—the Hittites, Amorites, Canaanites, Perizzites, Hivites and Jebusites—as the Lord your God has commanded you.**

Brief Solution To The Oddness

Here are my thoughts for whatever they are worth concerning death. People need to get the Holy Spirit "Light" or they cannot go beyond that Crown Chakra state that some of the Buddhist kind of people talk about AND they CANNOT enter the Universe of Heaven. That being said, the "common good" would be for as many people to get the Holy Spirit "Light" as possible and the only reason, I can think of for God waiting thousands of years to come back to the Earth to take his people home would be to get a higher percentage of people to go with him. Clearly, the percentage seems to be getting worse instead of better, so there is a strangeness to be reckoned. If the percentage is getting lower and lower, God would have returned sooner, so we can assume that the percentage is actually getting larger. That is where something we can call "re-entry of life" comes in.

Heaven War Losers

To study this important aspect of death, we need to re-examine this huge war that happened a long, long, long time ago. This is known as the Heaven War or War of the Titans and many other things. Massive numbers of people living in heaven revolted and tried to take control. After losing, the rebels experienced what could be called a second death. I think this second death thing will help us understand a primary death.

Turned Back Into Humans

Before the War, these people, known only as "watchers" by the Jewish texts, were different than us in that they had made the leap to live in this place we call Heaven. You can call the place alternate universe or linked universe or whatever. Heaven is just as good a name as any other. Because the only way we currently know to get and live there is through death, let's just say that these people had died and were turned into a new type of person who could live in Heaven. I would believe things were wonderful for a long time, but soon, a large group was established under the 3rd Archangel named Gadrael and he tried many things to take control. The Sumerians and other ancient texts talk about huge animals and dragons were created to help in the war, but even these massive monsters didn't give enough advantage to Gadrael and his troops. One third of the entire population of Heaven supposedly defeated hundreds of thousands of years ago and all the massive animals were destroyed, the Earth and its neighboring planets were turned into desolate places. The book of "Isaiah" tells us all the cities were destroyed in this war to end all wars. The rebels were punished in 3 terrible ways.

1. There were turned into "Normal Humans again
2. They were banished from Heaven forever and could never leave the earth
3. They had something only described as the "Light" taken from them so they could never attempt to take control again and they lost the capability to control reality no matter how much FAITH they had..

The losers whined a lot and then they whined some more. For thousands of years they were the rulers of the desolate Earth. After rebuilding it, they became rulers over all the people who had not been in the war. This was because they had knowledge they did not share and they lived a long, long, long time. Here are a few texts talking about this incident.

"Codex Junius II" [Germanic]

For the traitors to reward their work, he shaped a house of pain and laminations of hell—they boasted they would take the kingdom easily. The Lord high of heaven lifted his hand against their host. The erring spirits, in their sin did not prevail against the lord. In his wrath he smote their insolence and broke their pride and bespoiled his foes of bliss, he crushed his foes, drove out the rebels from their ancient home and seats of glory. Our lord expelled and <u>banished out of heaven</u> the presumptuous angel host. They suffer the pains and woe of tribulation, knowing bitter anguish. [One of the three punishments]

"Jubilees 2:9" [Essene Jew]

Nor may we take revenge on him because he has <u>stripped us of the "light"</u>. He marked out the borders of the world and created man in his own image with whom he hopes <u>again</u> to people heaven, with pure souls. [The Second Punishment. Please note that without the light, the Watchers could not take vengeance on any of the heavenly host. They lost some substantial power. Also, note that the word "again" is put in the verse to let us know that man was here before the war and was recreated after it was over.]

"Secrets of Enoch 31:4" [Gnostic Jew]
Because Adam was to be lord on Earth to rule and control it, the devil as a fugitive made War on heaven, thus he became different from the other Watchers. [This is talking about the rebels becoming like "normal Humans without this light thing.']

Gnostic "Light" Text
"Book of John the Evangelist"-"My father [God] changed his [Satan's] appearance because of his pride and the "light" was taken from him. His face became like a heated iron wholly like that of a man". [The old "heated iron man head" thing is not a very good description of living without this "Light" thing, but that is what the book said so I wrote it here.]

Isaiah 14:15
Yet thou [Satan] shalt be brought down to hell, to the sides of the pit. All they shall speak and say unto thee, Art thou also become weak as we? art thou become like unto us? [Here again we find that the leader of the rebel Watchers is now like a man. By losing the light, they became weak.]

Edger Cayce [20th Century Seer]
God moved and said,"Let there be light", and there was light. Not the light of the sun, but rather the light which—through which—in which—every soul had, and has and every had, its being. [While this verse comes from Genesis, it should be known that God said let there be light a long time before the Sun was put in its place, so it is talking about something else. Possibly, it is talking about the stuff taken away from the rebel watchers.]

Can Never Leave Or Control Reality
There punished people did eventually die, but they could never leave the Earth. More than that, these people "without the light" seemed to not be able to modify reality as Jesus told his disciples

to do. Stuck here forever in sort of a ghost state, these now dead people became known as demons, or worse. I'm not going to really get into this type of dying, because it is way too depressing. All I will say is that these roaming people still have not been allowed to leave so don't be surprised if things seem weird at times and please do not believe there are no demons.

Demons

When I talk about demons, I don't mean that all demons try to take control of people and turn them into mad killers or whatever. Some tried to be nice.

In Greece and India, we find that the aberrant teaching ANAK were called demons, not because they were thought of as bad but because they were the teachers of early man. To them, the word demon did not have a negative connotation. In fact, the Greek word demon actually meant, "knowing ones". Here are some of the things written about the ~~demons~~ knowing ones. It is evident that the ANAK had all died before these writings and now, they were spirits stuck here on Earth. Even after dying, some were still helping and sometimes harming man in some esoteric way.

Plato Describes Demons

Plato Wrote in "Dialogues, Laws" the following,

"There is a tradition of the happy life of mankind in the days when all things were spontaneous and abundant—In this manner God in his love of mankind placed over us the DEMONS, who are a superior race, and they, with great care and pleasure to themselves and no less to us, taking care of us and giving us peace and reverence and order and justice never failing, made the tribes of men happy and peaceful—for Cronus knew that no

human nature, invested with supreme power, is able to order human affairs and not overflow with insolence and wrong."

Hesiod Describes Demons

Like Plato, Hesiod wrote about the demons as the "men" who came before us.

"But now the fate has closed over this race, they are holy demons upon the Earth, beneficent averters of ills, guardians of mortal men."

These are the ways we should think about some of the ANAK [demons], but many became what we might consider the epitome of evil simply because it was so horrible not to embody a "living" person..

Up Until Now

Up until now, you probably have been skeptical of the idea of live or undead demons existing. Even if you think they might have lived in the distant past you are resistant to thinking there is anything to it today. While that is a nice way to think about demons, there is no history, text, science, or religious testimony that will make you feel better. The ANAK Demons are still with us just like souls cannot die, demons cannot die either and they cannot leave our reality..

In a nutshell, ANAK and their offspring ruled just about every part of the world until about 3 thousand years ago when they simply died out. That wasn't the end as they became something we lightly call demons as they could never leave the confines of the earth, even after death. The stories of their control, power, sometimes evil nature, and miraculous capabilities lived on well after they finally died.

For this search, we will have to look at historical documents, ancient discoveries, mythological details, religious testimony, and things terms ANOMOLY. The ancient demons were not anomalous, but the evidence of their existence has been treated as an anomaly by many not wanting to accept the obvious. Like the new vampire stories, there is a religious part to all of this. The religious details along with science and discovery will prove that ancient rulers of mankind were powerful, had vast knowledge and they existed on the earth for thousands and thousands of years. To give you an idea of them let me reintroduce Lilith.

Demon Making Lilith

Ancient Jewish texts tell us a lot. They developed a whole demon culture around the belief that Lilith was an evil and powerful being.

Bacharach, 'Emeq ha Melekh, 19c
And behold, the harsh husk, that is Lilith, is always in the sheet of the bed of a man and a woman who copulate, in order to take the sparks of the drops of seed which go waste, because it cannot be without this, and <u>she creates from them demons, spirits and Lilin</u>. And there is an incantation to drive away Lilith from the bed and to bring forth holy souls, which is mentioned in the holy Zohar.

Bacharach, 'Emeq ha Melekh, 102c-103d-
The Alien Woman is Lilith, and she is the sweetness of sin and the evil tongue. And from the lips of the Alien Woman honey flows. And although the Impure Female has no hands and feet for copulation, for the feet of the serpent were cut off, nevertheless the Female in her adornments looks as if she had hands and feet. And it is the mystery of her adornments that she can seduce men.... And she leaves the husband of her youth [Samael] and descends and fornicates with men who sleep below in the impurity of spontaneous emission, and from them are born demons and spirits and Lilin, and they are called the Sons of Man.

[The hands and feet reference is the curse that forced Lilith and her cohorts to stay on the earth even after death. There was no escape. They were ever present demons.]

I know all of the Lilith stories are confusing, but ancient historians thought that she was very important to the ancients as she was a very powerful entity, especially after she died. She became the leader of many of the demons. Some of them showed up during the time of Jesus, God incarnate.

Jesus Removes Demons

Whenever demons take on an entity with the light [that element that allows fro Anthropic reality control], they loved it and sometimes would make their host do things that seemed impossible with our NORMAL view of reality. Let's review those pigs again. While there are many references to evil spirits taking control of people, the Bible is a good reference of the misery of the demons. In this section, Jesus takes out a demon in Capernium.

Mark 1:23-26

And there was in their synagogue a man with an unclean spirit; and he cried out, saying, Let us alone; what have we to do with thee, thou Jesus of Nazareth? Art thou come to destroy us? I know thee who thou art, the Holy One of God. And Jesus rebuked him, saying, Hold thy peace, and come out of him. And when the unclean spirit had torn him, and cried with a loud voice, he came out of him. [Similar event Luke 4]

In this section, the demons run to pigs just to have a body.

Mark 5:1-14
And they came over unto the other side of the sea, into the country of the Gadarenes and when he was come out of the ship, immediately there met him out of the tombs a man with an unclean spirit, Who had his dwelling among the tombs; and no man could bind him, no, not with chains: Because that he had been often bound with fetters and chains, and the chains had been plucked asunder by him, and the fetters broken in pieces: neither could any man tame him. And always, night and day, he was in the mountains, and in the tombs, crying, and cutting himself with stones. But when he saw Jesus afar off, he ran and worshipped him, And cried with a loud voice, and said, What have I to do with thee, Jesus, thou Son of the most high God? I adjure thee by God, that thou torment me not. For he said unto him, Come out of the man, thou unclean spirit. And he asked him, What is thy name? And he answered, saying, <u>My name is Legion: for we are many</u>. And he began to implore Him earnestly not to send them out of the country. Now there was a large herd of swine feeding nearby on the mountain. The demons implored Him, saying, "Send us into the swine so that we may enter them." Jesus gave them permission. And coming out, the unclean spirits entered the swine; and the herd rushed down the steep bank into the sea, about two thousand of them; and they were drowned in the sea. Their herdsmen ran away and reported it in the city and in the country. And the people came to see what it was that had happened. [Matthew 8 and Luke 8 describe same event]

Here are a few of the other verses about eliminating possession of demons.

Matthew 9:33-34
And when the demon was driven out, the man who had been mute spoke. The crowd was amazed and said, "Nothing like this has ever been seen in Israel." But the Pharisees said, "It is by the prince of demons that he drives out demons."

Matthew 10:8
Heal the sick, raise the dead, cleanse those who have leprosy, drive out demons. Freely you have received, freely give.

Matthew 17:18
Jesus rebuked the demon, and it came out of the boy, and he was healed from that moment.

Mark 1:34
And Jesus healed many who had various diseases. He also drove out many demons, but he would not let the demons speak because they knew who he was.

Mark 1:39
So he traveled throughout Galilee, preaching in their synagogues and driving out demons.

Mark 3:15
And to have authority to drive out demons.

Mark 6:13
They drove out many demons and anointed many sick people with oil and healed them.

Mark 16:17
And these signs will accompany those who believe: In my name they will drive out demons; they will speak in new tongues;

Luke 4:33
In the synagogue there was a man possessed by a demon, an evil spirit. He cried out at the top of his voice, . . . Be quiet!" Jesus said sternly. "Come out of him!" Then the demon threw the man down before them all and came out without injuring him.

Luke 4:41
Moreover, demons came out of many people, shouting, "You are the Son of God!" But he rebuked them and would not allow them to speak, because they knew he was the Christ.

Luke 8:2
and also some women who had been cured of evil spirits and diseases: Mary from whom seven demons had come out;

Luke 8:35
and the people went out to see what had happened. When they came to Jesus, they found the man from whom the demons had

gone out, sitting at Jesus' feet, dressed and in his right mind; and they were afraid.

Luke 9:1
When Jesus had called the Twelve together, he gave them power and authority to drive out all demons and to cure diseases,

Luke 9:42
Even while the boy was coming, the demon threw him to the ground in a convulsion. But Jesus rebuked the evil spirit, healed the boy and gave him back to his father.

Luke 10:17
The seventy-two returned with joy and said, "Lord, even the demons submit to us in your name."

Luke 11:14
Jesus was driving out a demon that was mute. When the demon left, the man who had been mute spoke, and the crowd was amazed.

John 10:21
But others said, "These are not the sayings of a man possessed by a demon. Can a demon open the eyes of the blind?"

1 Corinthians 10:20
No, but the sacrifices of pagans are offered to demons, not to God, and I do not want you to be participants with demons. . . . You cannot drink the cup of the Lord and the cup of demons too; you cannot have a part in both the Lord's table and the table of demons.

1 Timothy 4:1
The Spirit clearly says that in later times some will abandon the faith and **follow deceiving spirits and things taught by demons**. *[Thins is an important element as demons who take control of a person can instruct he or she to inform or misinform. Additionally*

when our soul is separated from a carnal self, the Demon souls may also have sway.]

James 2:19
You believe that there is one God. Good! Even the demons believe that--and shudder.

Revelation 9:20
The rest of mankind that were not killed by these plagues still did not repent of the work of their hands; they did not stop worshipping demons, and idols of gold, silver, bronze, stone and wood -- idols that cannot see or hear or walk.

Revelation 16:14
They are spirits of demons performing miraculous signs, and they go out to the kings of the whole world, to gather them for the battle on the great day of God Almighty.

Revelation 18:2
With a mighty voice he shouted: "Fallen! Fallen is Babylon the Great! She has become a home for demons and a haunt for every evil spirit, a haunt for every unclean and detestable bird.

The main thing here to understand is that there were thousands of possessed people who had demons take control of them and these same demons have NEVER left our planet. Sure exorcists have been able to yank them out so they can find another pig "so to speak", but many times this was not a successful venture.

Jews Try To Drive Out Demons

In this case, people saw the Disciple of Jesus named Paul yanking out these vampiristic demons and some decided they could do that. They were wrong!!!

ACTS 9:13-16
Some Jews who went around driving out evil spirits tried to invoke the name of the Lord Jesus over those who were demon-possessed. They would say, "In the name of the Jesus whom Paul preaches, I command you to come out." Seven sons of Sceva, a Jewish chief priest, were doing this. One day the evil spirit answered them, "Jesus I know, and Paul I know about, but who are you?" Then the man who had the evil spirit jumped on them and overpowered them all. He gave them such a beating that they ran out of the house naked and bleeding

Don't try this at home or you may get beat up and thrown into the streets naked as well.

This section is put in the book to show you how very real wandering souls really are. If they are not comforted with a spirit, [either the one they had during life or the substitute Holy Spirit mentioned in the Bible], or stay asleep, or allowed to reincarnate another "self". There can be misery in life after life. To show that many times souls are reactivated or are reanimated, let's look at some of the people who were brought back 2000 years ago.

Raising the Dead

The Bible is filled with details about people raising the dead. Here are a few.

1 Kings 17:17-24
- And it came to pass after these things, that the son of the woman, the mistress of the house, fell sick; and his sickness was so sore, that there was no breath left in him. --- [Elijah] he took him out of her bosom, and carried him up into a loft, where he abode, --And he stretched himself upon the child three times, and cried unto the LORD, --And the LORD heard the voice of Elijah; and the soul of the child came into him again, and he revived. [Elijah raised the dead]

2 Kings 4:25-35
- The Shunammite woman --conceived and gave birth to a son.-- something disastrous happens to her son and he dies. The Shunammite woman hastily went to the prophet Elisha, [who] stretched himself upon him: and the child sneezed seven times, and the child opened his eyes.

Kings 13:21
- And it came to pass, as they were burying a man, that, behold, they spied a band of men; and they cast the man into the sepulcher of Elisha: and when the man was let down, and touched the bones of Elisha, he revived, and stood up on his feet.

Jonah 2:1-6-
[Jonah said after coming out of a whale that was underwater for 3 days] From the <u>depths of the grave</u> I called for help, and you listened to my cry. --I have been banished from your sight; yet I will look again toward your holy temple.--- To the roots of the mountains I sank down; the Earth beneath <u>barred me in forever</u>. But <u>you brought my life up from Sheol.</u>

Luke 7:13-15
- the Lord --touched the bier[dead boy]: and they that bare him stood still. And he said, Young man, I say unto thee, Arise. And he that was dead sat up, and began to speak. And he delivered him to his mother.

Matthew 9:25/Mark5:42/ Luke 8:55
- But when the people were put forth, he went in, and took her [a dead girl] by the hand, and the maid arose.

John 11:43-44
-And when he thus had spoken, he cried with a loud voice, Lazarus, come forth. And he that was dead came forth, bound hand and foot with graveclothes: and his face was bound about with a napkin. Jesus saith unto them, Loose him, and let him go.

Matthew 27:52-53
- And the graves were opened; and many bodies of the saints which slept arose, And came out of the graves after his resurrection, and went into the holy city, and appeared unto many.

Acts 9:36-42
- Now there was at Joppa a certain disciple named Tabitha, - it came to pass in those days, that she was sick, and died-- Peter put them all forth, and kneeled down, and prayed; and turning him to the body said, Tabitha, arise. And she opened her eyes: and when she saw Peter, she sat up.

Acts 20:9-12

a certain young man named Eutychus, being fallen into a deep sleep --fell down from the third loft, and was taken up dead. And Paul went down, and fell on him, and embracing him said, Trouble not yourselves; for his life is in him. And they brought the young man alive, and were not a little comforted.

2 Corinthians 12:1-4

I [Paul talking about himself dying the 1st time] know a person in Christ who fourteen years ago was caught up to the third heaven [Died]. Whether it was in the body or out of the body I do not know - God knows. this person was caught up to paradise.

Acts 14:19-20

Jews from Antioch and Iconium, -- stoned Paul, drew him out of the city, supposing he had been dead. As the disciples stood round about him, he rose up, and came into the city: and the next day he departed with Barnabas to Derbe [50 miles away]. [Paul was probably dead a 2nd time as people simply left his carcass, then he hopped up and walked 50 miles. He died a third time in Rome.]

While many people were raised from the dead, other people in the Bible such as Enoch, Elijah, and Melchizedek did not even die at all. They were translated. Certainly, their "self" would have been thrown away, but it must have been a very fast transition.

Sleeping Dead

The Bible is filled with descriptions of our soul "sleeping" after death, at least, for a while.

1 Samuel 28:11-15

Then the woman asked, "Whom shall I bring up for you?"- "Bring up Samuel," he said. When the woman [witch of Endor] saw Samuel, she cried out at the top of her voice and said to Saul, "Why have you deceived me? You are Saul!" The king said to her, "Don't be afraid. What do you see?" The woman said, "I see a ghostly figure coming up out of the earth." "What does he look like?" he asked. "An old man wearing a robe is coming up," she said. Then Saul knew it was Samuel, and he bowed down and prostrated himself with his face to the ground. Samuel said to Saul, "Why <u>have you disturbed me</u> by bringing me up?" Samuel had been "sleeping" and the Witch of Endor had awakened him. He was plenty mad about the whole thing. Certainly, he was not in Heaven and he had not been conscious until Saul had him summonsed.

2 Corinthians 5:8

We are confident, I say, and willing rather to be absent from the body, and to be present with the Lord.

Luke 23:42-43

And he was saying, "Jesus, remember me when You come in Your kingdom!" And He said to him, "Truly I say to you, today you shall be with Me in Paradise." Certainly, Jesus did not go to heaven right away and I can tell you it was not a lie. I'm not getting into a long dissertation about what a day is when you are dead, but if you are sleeping, day would be when you wake up so we can tell that the thief, was sleeping after he died until Jesus comes back.

John 11:11-44
He said, and after that He said to them, "Our friend Lazarus sleeps, but I go that I may wake him up." Then His disciples said, "Lord, if he sleeps he will get well." However, Jesus spoke of his death, but they thought that He was speaking about taking rest in sleep. Then Jesus said to them plainly, "Lazarus is dead. --- Jesus, again groaning in Himself, came to the tomb. --- He cried with a loud voice, "Lazarus, come forth!" And he who had died came out bound hand and foot with grave clothes, and his face was wrapped with a cloth. Jesus said to them, "Loose him, and let him go." Lazarus could not tell people about the afterlife, because he had not lived one. He had been "as if asleep". Jesus told everyone that death was like sleep. I don't mean a dreaming sleep. I mean a nothingness sleep.

1 Corinthians 15:40-44
- Behold, I tell you a secret: we shall not all sleep, but we shall ALL BE CHANGED, IN, MOMENT, in the twinkling of an eye, AT THE LAST TRUMP: for the trumpet shall sound, and the dead shall be raised incorruptible, and we shall be changed. Again, no question about everyone "sleeping" when they are dead until Jesus comes back. Everyone who is dead "in Christ" get yanked out of their graves and become immortal. The idea that everyone stays asleep in their graves until Jesus comes back is reinforced in "John", "Ecclesiastes" and "Daniel.

John 5:28
The hour is coming, in which all who are in the graves shall hear his voice, and shall come forth"

Daniel 12:2
Them that sleep in the dust of the earth.

Ecclesiastes 9:5
The soul, the living know that they shall die, but the <u>dead know not anything</u>. They are asleep.

Matthew 27:52
"The graves were open and many bodies of the saints <u>who had fallen asleep were raised</u>." This verse talks about when Jesus revived himself after his cross death. Not only does it help expand the confirmation of something miraculous happening, it also shows that people are still in their graves after death---"sleeping". Please don't think that they were yanked out of Heaven to re-inhabit their bodies. Peter reinforces the idea some more in Acts by bringing up David, still in his tomb and not in heaven.

Acts 2:29-34
*Men and brother, let me speak freely to you, the patriarch David, he is <u>both dead and buried and his tomb is with us to this day</u>.--- For **David <u>did not ascend into the Heavens</u>**."* Man oh man! "Why be Christian?" some would say, but Paul states it in the best way. "When Jesus comes back all the dead people rise and BEGIN their eternal COMFORT".

1 Thessalonians 4:15
-This we say unto you by the word of the Lord, that we which are alive and remain unto the coming of the Lord shall not prevent them which are asleep, because the Lord Himself shall descend from Heaven with a shout with a voice of the archangel and with the trumpet of God, see the last trumpet, happens with the second coming, the <u>dead in Christ shall rise first</u> and we which are alive and remain shall be caught up together with them in the clouds to meet the Lord in the air so shall we ever be with the Lord comfort one another with these words.

While people can be awakened from death-sleep, generally speaking, dead people sleep until Jesus comes back. To someone who has passed away, there is no sense of time as we understand it as indicated in Luke. While a number of people have clearly been brought back Samuel, Moses, Isaiah, Elijah, Jonah, Job and others there is a strong belief that very few dead people are "disturbed". Some believe there is a larger number of people who "return", but that is a different story.

Do We Know People After Death?

Let's look at one other thing about memory after death. For this, I am presenting a misconception in resurrection in the end of days that many people get confused when it comes to living whenever we believe DEATH has occurred. Jesus indicated in the book of Luke that all who follow him will never be married, will not die again, and are like angels.

Luke 20:27-30
-Jesus replied, "-- those who are considered worthy of taking part in <u>that age</u> and in the resurrection from the dead <u>will neither marry</u> nor be given in marriage, and <u>they can no longer die</u>; for they are <u>like the angels</u>. They are God's children, since they are children of the resurrection.

The issue, of course is about those who are married in this lifetime. They will no longer be married and there may be an issue about being angels that people once ignored. I would like to stay silent on this section but the recent loss of my daughter has made me think about this whole issue.

If you can't know your loved ones after you die, why is there a heaven and what good is it to ever be reincarnated? To start with here are a few of the texts to consider.

The short answer is we don't know our loved ones when our Soul is "released".

Jesus Resurrection

Here is what we do know, when Jesus first came back from the dead, no one recognized him. His voice was different, his looks were different. He had the same soul, but the physical body had been changed. Luckily, Jesus knew who they were and could relate to them, but he was in a brand new body. Very quickly, those who had not known him came to be close to him.

Isaiah 65:16-17
"For the past troubles will be forgotten and hidden from my eyes. For, behold, I create new heavens and a new earth: and the former shall not be remembered, nor come into mind."

Revelation 21:4
"He will wipe every tear from their eyes. There will be no more death or mourning or crying or pain, for the old order of things has passed away."

1 Corinthians 15:35–38-
A mortal body is like the seed, while an immortal body is like the full-grown plant. Both are physical, with an intrinsic continuity between the two. But what a difference between the seed and the plant in appearance, in attribute, and in potential!

This is pretty clear, while we may not remember exact people there will be an INTRINSIC CONTINUITY. There will be "things" we DO remember that help us and the re will be "things" about us other souls would remember.

You won't know your wife, child, friend, or acquaintance as they are today according to the Bible and other ancient texts. That being said, I am certain we will all know those we are empathetic with on earth. We might not recognize all the bad times and even the good ones, but I believe there is still a close union that is afforded interactions of people's souls over time. This may only be some "feeling" we will recognize, but I know that we did not have to go through this life only to be alone in Heaven. In our carnal world, we simply have no understanding about the interaction of each other's souls. Someday, I will know and possibly rewrite this section.

While there will be a level of shielding in our memories, there will be warmth and happiness and some type of union between the others living in that place called Heaven. This also means <u>the people we have encountered here on earth will be "known" in a way different than we know people today</u>.

By the way, don't believe those trying to tell you that because Lazarus knew the rich man, we will know our friends in heaven. That was not describing heaven at all. Also, some try to say because Moses and Elijah were re-animated in a similar body, we will have to have a similar body in heaven. That also is not a truth. While the truth seems bad to us in a way, the truth will be more pleasant that we can imagine. We will find our loved ones one way or another and have a deeper life together.

Dead Souls?

The Bible states that when we die the **"spirit"** is taken by God and the **"self"** becomes dust again. The discussions in the New Testament of John the Baptist being Isaiah and reincarnations of Moses, Elijah, Lazarus, many saints that rose when Jesus rose from the dead and all the rest are all showing our soul, the main part of us, doesn't die.

*Ecclesiastes 12: 7- "Then, shall the dust **[dead body]** return to the earth as it was, and the spirit shall return unto God who gave it."*

This is talking about dying and saying that when someone dies, his physical part turns to dust and his spirit, [sort of a life force] is completely separated from it. A person is only made up of a "self" and "Spirit" and a "Soul". The "soul" part is the part that doesn't die. Unfortunately, this is not always the case.

Ezekiel 18:2

-The soul that sinneth, it shall die; [As we have been discussing, just because you can live in death, it is not a sure thing.]

Psalm 78:50-

He spared not their soul from death; [This is specifically talking about angels that became human well before the flood of Noah. They did lose the ability to have a soul live after death and the Bible goes into detail about how Hell was made for this band of renegades.]

Lose Life to Gain It

Mark 8:31-9:6

For <u>whoever wants to save his life will lose it</u>, but whoever loses his life for me and for the gospel will save it. <u>What good is it for a man to gain the whole world, yet forfeit his soul?</u> -- "I tell you the truth, some who are standing here <u>will not taste death</u> before they see the kingdom of God come with power." After six days Jesus -- led them up a high mountain, where they were all alone. There he was transfigured --And there appeared before them <u>Elijah and Moses</u>, who were talking with Jesus.

I mentioned before that saving your life to lose it sounds odd and the idea that one could or could not lose his soul perks up your ears, but then there is a very strange thing stated here. It sounds like the writer is saying someone "will not taste death" for 2000 years before God comes again in power. Finally, Elijah and Moses are both reincarnated to help bring light to what was being said. Matthew and Luke both indicate almost the exact thing, because this concept was so important.

Matthew 16:24-17

For <u>whoever wants to save their life will lose it</u>, <u>but whoever loses their life for me will find it</u>. <u>What good will it be for someone to gain the whole world, yet forfeit their soul?</u>--- Or what can anyone give in exchange for their soul? For the <u>Son of Man is going to come in his Father's glory with his angels, and then he will reward each person</u> according to what they have done. "Truly I tell you, some who are standing here <u>will not taste death</u> before they see the Son of Man coming in his kingdom."-- After six days Jesus -- led them up a high mountain by themselves. There he was transfigured -- Just then there appeared before them <u>Moses and Elijah</u>, talking with Jesus.

Luke 9:23-33

For <u>whoever wants to save their life will lose it</u>, but whoever loses their life for me will save it. What good is it for someone to gain the whole world, and yet <u>forfeit their very self</u>? --- "Truly I tell you, some who are standing here <u>will not taste death</u> before they see the kingdom of God." About eight days after Jesus -- took Peter, John and James -- and went up onto a mountain to pray. As he was praying, the appearance of his face changed, and his clothes became as bright as a flash of lightning. Two men, <u>Moses and Elijah</u>, appeared in glorious splendor, talking with Jesus. They spoke about his departure, which he was about to bring to fulfillment at Jerusalem.

Please notice one thing here, when God comes in power is defined as the time Jesus and his angels return and judge everyone for the things they did during their lives. Let me try to paraphrase if I can.

- *Jesus said if you want your "soul to live", don't worry about your "carnal life".*
- *He reiterated his command a second time indicating that if one focuses on success in this world, he would forfeit his soul.*
- *He reiterated it a third time saying there is nothing that we have as precious as our living soul.*
- *He reiterated it a 4th time saying some souls will still be alive [and conscious] when I return in 2000 years.*
- *He emphasized a 5th time by having the living souls of Elijah and Moses come back to show these guys that their souls had not died."*

What a Contradiction!

Please notice that life and soul are sometimes interchangeable this is why he said saving your life loses your life. The second life is the life of the SOUL [most important part of you]. Let me give some more detail from John.

John 6:38-65

- this is the will of him who sent me, that <u>I shall lose none of all those he has given me, but raise them up at the last day</u>. For my Father's will is that everyone who looks to the Son and believes in him <u>shall have eternal life</u>, and I will raise them up <u>at the last day</u>.--- No one can come to me unless the Father who sent me draws them, and I will raise them up at the last day. They will all be taught by God. Everyone who has heard the Father and learned from him comes to me. --I tell you, the one who believes has eternal life. -- Whoever eats my flesh and drinks my blood has eternal life, and I will raise them up at the last day. For my flesh is <u>real food</u> and my blood is <u>real drink</u>. ----The <u>Spirit gives life [Soul]</u>; the <u>flesh[self] counts for nothing</u>. The words I have spoken to you—they are full of the Spirit and life [soul].

Main Theme

At death, the "spirit" portion of a person returns to God. The Body turns to dust and dies away. That leaves the soul which can either die or have eternal life. The part can be reincarnated even if the body has turned to dust. It is the part that allows some of God's disciples to "NOT SEE DEATH UNTIL HE RETURNS"

When dead, the SOUL is, typically, in a completely unconscious state, but it can be reanimated into someone similar to the original [as the Reanimation of the "saints" when Jesus rose or Moses and Elijah when Jesus was transfixed or Lazarus after his death] or Reincarnated into a completely different individual that has almost no knowledge of his previous life [as John the Baptist, the King of Tyre, or similar reincarnation] or one can be brought back [as a ghostly figure as Samuel when revived by the Witch of Endor. Let's read more to confirm this.

John 8:51-52

*-[Jesus said,] "If anyone keeps My word he[**his soul**] will never see death." The Jews said to Him, "-- Abraham[**carnal body**] died, and the prophets also; and You say, "If anyone keeps My word, he **[soul again]** will never taste of death."*

2 Corinthians 5:6-8
-Therefore we are always confident, knowing that, while we [__our souls__] are at home in the body, we are absent from the Lord: For we walk by faith, not by sight. We [__our souls again__] are willing to be absent from the body, and to be present with the Lord.

1 Thessalonians 4:15
-For this we say unto you by the word of the Lord that we [__soul part again__] that are alive, that are left unto the coming of the Lord, shall in no wise precede them [__souls__] that are fallen asleep. – the [__souls of the__] dead in Christ shall rise first; then we [__Soul part__] that are alive that are left, shall together with them be caught up in the clouds, to meet the Lord -So shall we [__our Soul with glorified, noncarnal body__] ever be with the Lord.

Matthew 16:28
-Some who are standing here [__their souls__] will not taste death before they see the Son of Man coming in his kingdom.

Matthew 24:34
This generation [__of souls__] will certainly not pass away [__die due to disbelief__] until all these things have happened. John 11:26- whoever lives by believing in me [__then their soul__] will never die.

Hopefully that is as clear as mud. I have been trying throughout this book to describe this dilemma and I want to make sure that, if you don't get anything else out of this book you understand that self centered pleasure will surely lead to misery. Quit it right now. Just quit. Go out and help someone, truly learn empathetic love, and humble yourself and you will be happier, and experience a life beyond your dreams. I know this part has been religious, so I might as well go ahead and say---- "If you want to experience this same power, warmth, and love beyond the second coming of God, focus on God incarnate.--- OK that's all the preaching I'm doing. Let's get back to a section in the Bible that clearly describes a reincarnation through killing.

"Good" Disasters Show Reincarnation

Romans 8:28 seems to be saying God, who brings hurricanes and disasters which kill thousands of people every year, is doing these things for the good of people who LOVE God. Initially, this sounds weird, but let's look deeper. By the way, the idea that God does not have control over these things is not the answer. If he is not omnipotent, omnipresent, and omniscient, there would be other, more serious issues with the Biblical details.

Romans 8:28
And we know that <u>all things work together for good</u> to them that love God, to them who are the called according to his purpose. Talking about the various faithful ancestors of the Jewish faith, we find the following in Hebrews and Ecclesiastes that gives us half of the answer.

Hebrews 11:13-12:11-
-These all [all the Jewish patriarchs] died in faith, not having received the things promised, but having seen them and greeted them from afar, and having acknowledged that they were <u>strangers and exiles on the earth</u>. -- Women <u>received back their dead by resurrection</u>. Some were tortured, -- - They were stoned, they were sawn in two, they were killed with the sword. "My son, do not regard lightly the <u>discipline of the Lord</u>, nor be weary when reproved by him. For the <u>Lord disciplines the one he loves, and chastises every son whom he receives</u>." It is for discipline that you have to endure. God is treating you as sons. For what son is there whom his father does not discipline? They

disciplined us for a short time as it seemed best to them, but he disciplines us for our good, that we may share his holiness. For the moment all <u>discipline seems painful</u> rather than pleasant, but <u>later it yields the peaceful fruit of righteousness</u> to those who have been trained by it.

Ecclesiastes 7:14
In the day of prosperity be joyful, and <u>in the day of adversity consider: God has made the one as well as the other</u>, so that man may not find out anything that will be after him. The second part of the question comes back to the living soul. Corinthians tells us a truth.

1 Corinthians 15:51
*-Listen, I tell you a mystery: We will not **all** sleep, but we will all be changed.---* <u>We [the Soul part of us] will not ALL sleep, but ALL [souls] will be changed [whenever Jesus returns].</u> What this is saying is that some of us will be able to have our souls wake up after death. That would be useless unless we are reincarnated, resurrected, or revitalized in some way. You can forget the resurrected before Jesus comes back, so SOME people do come back.

Die and Get a Second Chance
If someone is given a second chance at accepting God, the whole idea of him losing his life in a hurricane is a GODSEND. I know some are going to balk at the idea that God allows some to have a second chance, but the details are provided over and over. One can say that the only people that are returned to life are those who accept Jesus, but that just isn't so. Many of those who were revived to life by Jesus had never been his followers, yet they were allowed to have a second chance.

When God forced the Jews to kill every living person in towns like Jericho and the cities of the Canaanites, certainly it was done for the good of the Jews. At the same time, if unbelievers were given a second chance by reincarnation, the destruction and killings were a GODSEND for those who were destroyed. Let's say since Adam there have been something like 100 quintillion [1 with 15 zeros or 30 million times as many people compared to the number currently living on Earth] people give or take a thousand and let's say of those a couple billion have a strong acceptance of Jesus and can be taken into the clouds at the end of days. The idea that God forces the other 100 quintillion to be burned in the fire made for the rebel angels is not exactly what is presented in the Bible. Instead, we know that many have been brought back to life for a "second chance". There is a high probability that the number of people brought back is VERY high making the 100 quintillion number not the number of souls but only a much smaller quantity that allows for measly habitation of the 3 trillion people living in the world today. Let's look at some more texts.

Jude 1:4-23

*For certain persons have crept in unnoticed, those who were **long beforehand marked** out for this condemnation [during and earlier life], ungodly persons who turn the grace of our God into licentiousness and deny our only Master and Lord, Jesus Christ. Now I desire to remind you, though you know all things once for all, that the Lord, after saving a people out of the land of Egypt, subsequently destroyed those who did not believe. ---that they were saying to you, "**In the last time** there will be mockers, following after their own ungodly lusts." **These are the ones who cause divisions**, worldly-minded, devoid of the Spirit.....keep yourselves in the love of God, waiting anxiously for the mercy of our Lord Jesus Christ to eternal life. And have mercy on **some**, who are doubting; save others, snatching them out of the fire; and on **some** have mercy with fear, hating even the garment polluted by the flesh.* Even some of the really hateful people God will find a way. It specifically says that some of these people crept BACK INTO EXISTANCE who were LONG BEFOREHAND MARKED. What could this mean? To show that there are chances, Jude tells us emphatically, that the real problem people will be those souls who are animated in the "Last days" which cause derision. He does not indicate that the horrible people before that are the problems as they could get another chance. Notice in 2 places the word SOME indicates that people are not guaranteed a second chance.

I Timothy 2:3
*This is good and acceptable in the sight of God our Savior, who **desires all men to be saved** and to come to the knowledge of the truth.* If God wants all to be saved, and we have proof that some were given a second chance, and we have the odd nature of God forcing the Jews to kill every unbelieving Canaanite sounds odd when reading this verse.

2 Timothy 2:25
Opponents must be gently instructed, in the hope that God will grant them repentance leading them to a knowledge of the truth Again we find that God is longing to allow repentance to all.

John 3:17-
For God did not send his Son into the world to condemn the world, but to save the world through him. Notice it does not say "save a small number in the world", but as many as possible. After allowing some a second chance, it seems practical that others also had this chance.

1Timothy 4:10
*-That is why we labor and strive, because we have put our hope in the living God, **who is the Savior of all people, and especially of those who believe.*** If there was any doubt, Paul writes that Jesus is the savior of all, then adds the "especially" but there is no doubt that Jesus would provide a means for those who did not currently believe.

OK! I'm sure you have strong feeling about this discussion, but the reason for the overview was to investigate the act of killing "innocent people" in disasters for the good of believers and the requirement of God to "kill every single Canaanite" that the Jews saw, for the good of the believers. It also will help us investigate why it was better for Judas to have never been born [or at least born into that time].

Re-Entry of Life

I know some call this reincarnation, but I wanted to be different. As I showed earlier, Jesus' Disciples believed in this characterization during the time they were being taught by Jesus himself, so one might believe that some form of reincarnation is possible. The strange discussions of the mysterious Melchezedek who somehow survived for thousands of years seem to indicate this characteristic and "John the Baptist" was thought to be reincarnation of one of the ancient prophets. Over and over again, this seemingly absurd reincarnation keeps popping up in the Bible and many other ancient religious texts but most try to ignore it because it is uncomfortable. One might even wonder how the preflood people lived hundreds or thousands of years. If sequential life cycles were possible, the time constraints go away and the Bible begins to make more sense.

Why is God Waiting 2000 Years to Come Back?
If you have never wondered why Jesus is waiting 2000 years before coming back to earth, I think something is wrong with you. That is a long time. He told his disciple that he would be coming back VERY soon, so there is a mystery. My belief is simple. God wants as many to come with him as possible. He wants as many to take in that Holy Spirit/LIGHT as possible and be able to go with him but he wants to use this free will stuff. Therefore, he has instituted a remarkable means to allowing for both and mankind gets the opportunity to learn great and wonderful truths as God waits for his return date. I know this sounds like some spiritualist is ranting the sayings of an ancient book and shouting "BELIEVE" a few times to convince, you, but as we get into the

physics of all of this, you will see that there is no other "logical" answer.----At least, I have not found one and I have been doing quite a bit of research on this subject so you wouldn't have to.

Man Is Not Naturally Good
The apostle Paul of the Bible indicated that --

> *"Man thinks only of evil continually."*

Apparently, that means that we stay in the lower chakra levels most of the time [Sex, Self centered, self awareness, and Survival]. I call it the three "Ss". Another way of saying this is that, most of the time; we are comforted by simple pleasures of carnal living. Unfortunately, I know that Paul was correct in my life and I'm sure that you must say the same thing. I would suppose that evil is in the eye of the doer but the idea is that man cannot attain conscious transfer to the spirit universe by himself. If you die before you gain that "help" you are in trouble.

Your free will allowed you to stray from the greater truth and troubles arise.

Talking to the Dead
While there are countless accounts of people talking to dead people that go along with the previous Biblical review of Samuel coming back to talk to Saul, we should, at least be open to the probability that our consciousness does not die with the DNA. Looking at the timing of God's return, his statements about killing people for their own good and it being better not to have been born suggest that after death our being may be subject to reentry into bodies---

---to give us another and another chance at understanding, accepting, and adjoining with God[This, of course is for a limited number of times that are not known to us and each

successive year, the earth seems more carnal and less favorable for a transformation.]

What I Mean By This

This is certainly going to sound Biblical, but people aren't good at being the type of people God needs us to be and the type of people we, ourselves need to be. Christians believe that only this Holy Ghost can change us and because of our being bad, only God Incarnate's sacrifice could allow us to get this Holy Ghost. As a similar religion, Eckankar almost says the same thing except they start with a premise that people can get better without this Holy Ghost by learning from "teachers" that instruct us on how to improve ourselves. If one gets good enough, they can then, sort of get this Holy Ghost thing on their own. All along the way, we learn more and more about the nuances of living in a carnal and a non-carnal world.

Many of the lessons we must learn require suffering.

Hopefully, you can sense that God causes and allows suffering to <u>HELP</u> us learn, grow, understand, develop, and fulfill the requirements of God. Most religions seem to have this as a common thought. However, they typically identify the "helper" as not being of God, but as being others who have become enlightened. For this overview, either type of helper will demonstrate the same thing.

God promotes good by allowing things we think of as evil exist.

Unfortunately, our free-will, many times, gets in the way of our learning and our accepting and the percentage of those who simply cannot do what is needed to accept God's Holy Spirit and the world of the carnal living gets more and more carnal **every year**. When God Incarnate returns, those who can follow him retreat as horrible times await those remaining in this world. Nostradamus, Mother Shipton, Estras, Daniel, John, and many others predicted this. The terrible times seem to occur as the consciousnesses of those who had taken the Holy Spirit that would not have the "stability" of interface with the spiritual universe. It is the beginnings of universal instability that initiates and fuels the tragedy identified in these predictions and the timing, by all common sense, is soon. That is called Hell. I will have to describe some of this affect later. Right now, only think of rejuvenating life.

Many Deaths

That was getting a little heavy so let's move away from here and just go into definitions. I think I had better figure out what death is in the general sense. Once we can establish that, our journey will become easy. Unfortunately, there seems to be a lot of different deaths and we may have to pick and choose.

- **Hell is death-** Not only is it defined as death, it is also determined to be everlasting death. I think I need to put that in a special place. While I will define it a little more than you are used to, I don't think I will get into it much as it is too scary for me.

- **Death where a traveler is transported to many hells and heavens-** While most seem to be rejuvenated into new bodies after death, sometimes, before a return, investigations into a wide assortment of places has been possible. Enoch and many others left us with some of the details.

- **Recurring death and reincarnation-** No! I'm not talking about coming back as a bug; however, I have known some people who came back as snakes. We will have to look at that type of death for sure.

- **Learning Transition-** On the same line as that above, many now have a belief that one must return to satisfy some mistake or yearning after death. I'll give you some of the details of this interesting thought process and some of the evidence.

- **Death and resurrection** in the future- as promised by the Biblical accounts of Daniel, Thessalonians and Revelation. This one certainly is interesting and worthy of our thoughts.

- **Death is the halting of body operations**. While this one is one many people believe, the chances that we become nothing after heart stops looks less and less like death, the more we investigate. Our investigation here will be limited to Near-death experience rather than the whole enchilada.

Reincarnation Versus Resurrection

Let's continue in this vein and look at some of the definitions of death. Rather than explaining, it might be better to simply list a few verses that can help here. Afterwards, I will try to put perspective on them.

Reincarnation

Isaiah 53
After the suffering of his soul, he will see the light [or take in the Holy Spirit thing] and be satisfied; by his knowledge my righteous servant will justify many, and he will bear their iniquities. [This is identifying everyone as being iniquitous, promising substantial suffering, presenting justification and understanding by God Incarnate, and finally this resurrection thing. Please notice that it is talking about the SOUL suffering rather than people suffering. Why do you suppose that is? I'll tell you. The Soul extends beyond one life.]

Isaiah 66
"From one New Moon [birth] to another and from one Sabbath to another, <u>all mankind</u> will come and bow down before me, says the LORD. And <u>they</u> will go out and look upon the dead bodies of those who rebelled against me; <u>their worm will not die</u>, nor will their fire be quenched, and they will be loathsome to all mankind." [We don't know what else Isaiah might have written for his book ends with this final warning. The rebels mentioned here seem include all those who would not bring in the Holy Spirit. This included those who died well before the time of Jesus. The only way all mankind worship God and view the dead rebels

is for most people to take this Holy Spirit in during one of several successive lifetimes allowing for the acceptance.]

I've brought out a number of these verses, so I won't belabor them.

Resurrection

Reincarnation is different than Resurrection, in that many reincarnations may occur before this final RESSURECTION. Thing that changes the characteristic of humans such that the Physical bodies now can continue just like the consciousness had done over and over again using reincarnation.

Isaiah 26
But your dead will live; their bodies will rise. You who dwell in the dust, wake up and shout for joy. Your dew is like the dew of the morning; the earth will give birth to her dead. [I know I brought this up previously, but some try to say this whole dust thing means that people stay in the dirt for thousands of years and finally you wake up and you are either feeling really, really good or really, really nasty. Please substitute dust with "Earth" or "this universe", or whatever you feel comfortable to get this out of your head. Your body will be turned into dust, but you will not and YOU WILL NOT SIMPLY STAY IN TH EGROUND.

John 11
"I know he will rise again in the resurrection at the last day."

Ezekiel 37
Therefore prophesy and say to them: 'This is what the Sovereign LORD says: O my people, I am going to open your graves and bring you up from them; I will bring you back to the land of Israel.

I Corinthians 15
So will it be with the resurrection of the dead. The body that is sown is perishable, it is raised imperishable.

These don't paint a picture of dying and going to heaven or hell. What they say is that at the end of time, there will be an accounting or "second Death". The first death or deaths are provided to us to allow us to learn and finally take hold of this light thing and other things that allow us to live after the second death time. I think the book of Daniel may help us more in this description.

Daniel and Death

Daniel 11 and 12

"Those who are wise will instruct many, though <u>for a time</u> they will fall by the sword or be burned or captured or plundered."

Please read this one several times!

People will be teaching others <u>even after</u> they have fallen by the sword.

"Some of the wise will stumble, so that <u>they may be refined, purified and made spotless until the time of the end,</u> for it will still come at the appointed time."

Please read this one several times!

EVEN the wise will "stumble" in one life, but they will have more chances until the end of time.

"Multitudes who sleep in the dust of the earth will awake: some to everlasting life, others to shame and everlasting contempt."

Please read this again

After many reincarnations, there will be a resurrection. If, after many stumbles, one finally takes the Holy Spirit, this resurrection will be pretty neat.

When someone talks about resurrection, it does not mean that your nasty body comes out of the ground to mix with your heavenly spirit that has been dangling around for thousands of years. All of these verses tell us the same thing. Most people die and did not do what is needed to live a life in the "other side" that is nice. One can try, but without that Holy Spirit that Jesus talked about, most fail. Coming back allows a retry, and another and another. Hopefully each time people get closer to God. The Amorites had to die for their own good and we must suffer with unpleasantness for our training.

Not a Free Ride

This is not a free ride and one cannot believe that doing bad in this life can be forgotten in the next. It doesn't work like that. Every year this world is modified to be more and more Carnal [that is less and less the way God created it to be.] It is harder to become integrated with God in each successive lifetime. That is why God will put an end to it in the near future when the percentage of those who will stay with him are the highest possible.

If this were not so, the thousands of years God stays away does not make sense. Jesus continued to tell his disciples that the time of God's return was at hand. They thought it was to be in their lifetime. In one way, it was as more and more believers are collected for his final return and we are not aware of any time spent between lives or in previous lives, so to us the time is still at hand.

The Holy Spirit described in the Bible and this thing called Light by many ancient documents and tales seem to be the same thing and it is the most vital thing we must capture to understand and live peacefully in whatever type of after life there can be.

How long our conscious minds remain in this universe is a mystery. If we are to believe the ancient Biblical teachings, one might believe all souls have stuck around for thousands of years waiting for this final exchange.

Death From No Heartbeat

Let's continue by examining the idea that when you die, all your thought, dreams, ideas, wants, experiences, advances with universal interactions and anything else you can think of simply disappears when one DIES. Sometimes this idea makes me mad, but today I'm OK so let's discuss it.

As we looked at in the first part of this book, today's models of the universe are characterized by 3 independent dimensional blocks. One creates things. Another creates forces that allow interaction of things, chemical attractions, magnetics, electricity, photons, Radio waves, etc. All that is interesting, but now we know that there is still a third three-dimensional characterization required to build a universe. This last component allows things to live, think, dream, die, and, most importantly allow for a carnal existence.

Dreaming

Go to sleep! Can you still see things? <u>Where does the light come from</u>? Where does the entire world come from? I know you are just saying imagination, but there is much more to it. There is actually a change in the world from you dreams. The changes are miniscule, but one could sense them if they had some type of meter to test that sort of thing. The light you see in your dream is just as real as the light you "sort of perceive" when your eyes receive electromagnetic waves that reflect off surfaces. That jumbled mess of vibrations going forward and backward and side to side and being absorbed and changing wavelengths is a poor excuse for something we call light. What you really are interested in is perception. Perception of a dream and perception of what you have defined for yourself from all the electromagnetic waves are still just perception. The reason I am bringing all this up in a death book is that death is also perceived. We might go farther and say even Hell is "perceived rather than being something we could call real.

Hellish Hell

As people gain that level of insight, they potentially can better accept the "Spirit of God" which allows for transfer to the other universe unrestricted. The cyclic nature of death and the idea that addition lives will bring more suffering are all REQUIREMENTS to allow us to develop. It would be a bad thing for God to simply allow us to die and almost all those entities to be separated from him forever in some type of torment. I'm not talking about a lake of fire here. I'm talking about something worse than we can imagine. While many don't like to accept it, God holds our universe together. Separation from him means the universe structure comes apart. Our reality would blur. Interaction with other conscious entities would become strained and finally, each consciousness would be alone, desperately trying to link up to others in this same existence. Without universal structure, simply holding one's self together as an entity would be a full time, unimaginably difficult job. Don't worry about fire in a Godless universe. Fire could not exist.

When I think of hell, I think of uncontrolled astral projections. As one removes himself from some level of universal order, it is not always a good thing. I know most want to stay away from the subject and those in many religious circles get so tightly wound around some dogmatic description that they cannot even read about other theories of its existence. Well! Here goes the author's determination.

Hell is real and Hell is not real.

"Man what a cop-out!" you say, but let me explain. The Bible tells us that Hell is complete separation from God. It indicates that those who fail to accept the salvation freely given by Jesus, God incarnate, and take in something called the "Hoy Spirit", another component of God, he or she will remain in this place of darkness, pain, suffering, blackness, ENTROPY, almost non-vibrating, possibility of life without life. Does that sound similar to what I have been saying? Possibly, we can gain a nugget from looking at hell. One must get help "holy spirit" to vibrate at a level that allows complete union with the Spirit universe of Heaven.

For those who think that everything in this universe simply works like a clock and continuously runs is stupid. ------ I'm sorry I said that and I take it back. Please forgive me. What I mean is that all of this in-wave and out-wave stuff is fine to model our existence, but the fact of the matter is that ENTROPY is one of those laws that seems to ACTUALLY be a law of this carnal universe. Without God as the watch winder, everything would soon lose its -----existence as it travels to its lowest vibrational state. Let's go back in history.

ANAK

As I have been trying to explain, after the first Heaven War identified in the Bible and over 20 other ancient texts we find a significant reference to understanding life. The Jewish texts indicate that the people who had been on the rebellion side lost their "Light" and that without this "light thing" these people [called ANAK or "those who have fallen"] were unable to return to Heaven and they didn't completely belong in this universe. Many simply ignore these statements or put some dogmatic saying around them, but if we sense light, not as electromagnetic vibrations but as higher vibrations of consciousness as depicted by the "near death experiences" and that getting to these very high vibrations requires the thing Jesus called "Holy Ghost or Holy Spirit" we could make a strange move and call the Holy Spirit "Light". This would not be the normal light for seeing, but the light that is used by us to examine the environment. The special environment that could be seen with this Holy Ghost was the modified environment associated with a special change in the universal resonance. It seems to be this "vision" is required to allow any extended entry into the adjacent universe.

That would be against some religious teachings, but there is a reason Jesus told his followers that he was the Light of the world. I think he was telling the truth. Just like people today gain the "holy spirit light" to allow expansion of vibrational levels; the ancient ANAK people lost it and were confined to the carnal earth. When these guys died, they could not leave the carnal world completely and entities we refer to as demons were born. These "demons" or spirit remains of the ANAK people are not necessarily monsters, but they are stuck here. There, evidently, is a horrible loneliness that the "demons" have without the light. One Biblical description had a large number of these guys trying to associate with some live pigs just to feel alive, but that was short lived as the pigs killed themselves and the demons were again in this nasty state between living and not living. If it is so bad thousands want to experience existence inside a pig, we are talking about agony over thousands of years.

Think of it this way, Hell is when you want to find a pig to live in and there is no such thing.

For the ANAK souls, life is not as easy as it is for us. Body death seems to be just a stepping stone to understanding which helps us begin to go beyond ourselves and become closer to the spiritual world. Many will not try to understand faith and self actualization. They will not understand the joy of extending the carnal life to assure reasonable life after life.

How Do People Die?

There is a strong possibility that today, as people give up their bodies in death, many cannot leave just like the ANAK because they never gained the light of the "Holy Ghost" and lost their own spirit for one reason or another. If that happens, there are only 3 things that can occur. One is that the person sleeps for a time. A second thing would be that the person is revitalized in a new body, still having partial memories of a past life and still having another chance to get the Light from the Holy Ghost.

A third thing may also happen. That is, the person may become "Like" one of the ANAK demons.

Again, I don't mean anything sinister by this. All I mean is that one may remain conscious even without his body. Most of the time, I believe one would be awaiting a new body to inhabit, but other times, there may be a separation that does not allow reentry and one would become as one of the demons. I don't begin to understand just how and when this horrible change may occur, but I think, there are only a certain number of chances that are afforded for one to gain the "light" so one should not take death lightly. No matter, about the various levels of death, there are ample observations that there are some invisible entities around us. They aren't quite here and they are not quite in the spirit universe. It would be sort of like something we call hell --- separated from God and not able to reenter the carnal world either.

Supposed Spirit Teachers

Those experimenting with astral projection and out of body communications with the "other world" could, possibly be trapped in this in between world if they are not careful. The teachers they meet may not be from the spirit universe, but instead, they may be from this in-between area. I'm sure they can be friendly, they can be helpful and they can bring insight, but there also could be entities that try to gain comfort in numbers by not allowing you to follow the "Light" to experiencing the true direction of what we could call the CROWN CHAKRA.

Therefore, the best way to gain LASTING higher vibrational existence, one needs to somehow link up with the characteristic called Holy Spirit. Maybe there are other ways to gain levels of high vibration but I think that is a good one.

After a Final Death

Of course, like everyone else, I don't completely know what happens after death, but there is a peculiar thing we recognize from Biblical texts that sometime we ignore.

In the book of Revelation, it states that *"when God returns to the Earth in glory, the dead in Christ will be raised"*. After "sleeping" for thousands of years, these people wake up and rise up into the air so there is a mystery here and it seems to go along with my previous statements.

What happens to people who are to be raised up? Some tell us that they are in a state of slumber of halted time. Some indicate that only bodies are raised up in the last days and the consciousness has already gone somewhere else [I know that doesn't make sense, but I just putting it out there.]

Here are my thoughts for whatever they are worth. People need to get the Holy Spirit "Light" or they cannot go beyond that Crown Chakra state and enter the spirit Universe of Heaven. Therefore, the common good would be for as many people to get the Holy Spirit "Light" as possible and the only reason, I can think of for God waiting thousands of years to come back to the Earth to take his people home would be to get a higher percentage of people to go with him. Clearly, the percentage seems to be getting worse instead of better, so there is a strangeness to be reckoned.

If the percentage is getting lower and lower, God would have returned sooner, so we can assume that the percentage is actually getting larger. That is where re-entry of life comes in.

While the soul lives and reenters, and sleeps, and advances throughout time, during the last days, the soul will either DIE or not. Having your soul die isn't the best thing, so think about it.

Your Soul is the real you and it is hard to kill your soul.

Conclusions

Light, life, and death have similarities and differences. In order to investigate death, we had to understand light and life a little. Hopefully you got the following out of this book.

- **Light is not Light**-We looked at how light was different than what has been imagined as light in the past. While the apparent affect of light can be initiated by placing stress on particles with electromagnetic [In-waves], the electromagnetic waves do not cause light. They cause electromagnetic fields. Light is established in our consciousness. To that end, the color red is completely different to all people. While we define it as the same, what might be a more vibrant red hue might be interpreted as less bright by another or even a completely different color by an animal seeing the same vibrations.

- **Electromagnetic vibrations-** don't light up anything. While they generally are present during feelings of light and visual comfort, sometimes, light can be sensed with our eyes closed or during near death experiences showing eyes are not the important part.

- **Sideways Light**- We discussed how light is transposed from normal life and if time were viewed sideways, the thing we call light would appear as a solid mass while life would simply have excursions where an entire lifetime could be viewed simultaneously.
- **Matter ain't matter**. Instead, it is made up of what Einstein described as undulating nothingness. I will call it vibrations. [The same can be said about what I call the operational dynamo made up of these in-wave things to make forces. The Ethereal dynamo is made up of these vibration things as well.

Everything has a similarity and nothing has an end.
- **Nothing has an End**-The important thing that many now profess including Einstein and Dr. Wolff is that matter never ends. It sort of fades away as the vibration ripples get father and farther apart away from this seed, sometimes called standing waves as the in and out-waves seem to cancel each other out at one point in their travels. It is this neutral point that could be considered Fermionic.
- **Vibrations Must be Replenished**-As vibrations are lost in out universe, they must be replenished from another universe. It is the nature of time in an adjacent universe that actually allows for sustainment. There is conservation of everything so anything leaving must be "somehow" coming back.
- **All Dimensional Components Vibrate**-These matter vibrations are moving in 3 mutually perpendicular ways. All dimensions that make up the universe vibrate including dimensions in Structural Dynamo [at make up stagnate matter], the Operational Dynamo [that make forces] and Ethereal dynamos [that produce conscious life]. Everything is similar.
- **All Dimensions are grouped in sets of 3 perpendicular components**-The Operation dynamo is the one we know the most about, because it causes motion. In our infinite wisdom we called one of the dimensions electric fields which we know to be perpendicular to another dimension we typically call magnetism which is mutually perpendicular to the third of the operational dimensions that sort of joins the characteristics of both to produce Electromagnetism and photons. Guess what!! The other 2 dynamos must work the same way.
- **Stagnate Matter is three dimensional**-As an example we know that anytime particles are produced, all of a sudden there is something we call gravity. Putting a vibrating fermion and gravity together is something we can loosely call nuclear attraction. All three of these things produce what we might think of as stagnant matter.
- **Life is three dimensional**-As another example, no one has a good definition of life. It certainly is not DNA, but most of the

ancient religious data tells us that a being is made up of three entities [consciousness/self, soul, and spirit] or [id, ego, Baa] or on and on we could go. These three ethereal dimensions act just like the other 2.

- **All Dimensional groups resonate-** We know that any electric and magnetic field have a particular vibrational quality that places them in something we call **resonance**. Resonance is when dimensional qualities place the least or the most stress on the universal structure. Dimensional duals try to vibrate at this resonance as it is provides the most stability. Guess what! The other 2 must have this same characteristic.

This is important ---LIFE RESONATES.

- **Dimensions take in and give off energy out of phase-** There is a certain capacitance and induction that is noted by 2 dimensions in a dynamo. While we know that to be a fact in the operational dynamo with electricity and magnetism and that both characterizations are phase shifted from the other so that it is not sensed by the other. ---Any dynamo should have the same characteristic. Capacity or capacitance is the capability to take in or secure energy and induction is the ability to provide or use energy. Because the dimensions are mutually perpendicular, the 2 linked dimensions of a dimension characterized by a capacitance cannot sense the capacitance just like Induction is hidden from the others.
- **All dimensional Energy Equations MUST be able to be presented in the same form.**-If one energy equation is represented by one dynamo against time. All energy formulae should be represented in that same way. This last one is the issue that I have with the $E=MC^2$. This happens to describe energy or in-ways with respect to one of the dimensions in the Particle dynamo and other formulae have a different structure. $E=1/2\ CV^2$ of $E=1/2\ LI^2$ which describe out-waves with respect to dimensions of the Operational dynamo.
- **Time is a dynamomic dimensional component that ties the**

three dynamos together in a mutually perpendicular way.- We can look at time as another dynamo made up of the three dimensional dynamos. It is sort of a super dimensional dynamo. Without any one of the other three, time would not be possible.
- **Aether, Electricity, and Carnal or base Life** are joined together as static charge or building characteristics of their associated dimensional dynamos.
- **Magnetism, Gravity, and human spirit** are joined together as the kinetic elements of the dimensional dynamos.
- **Life Not DNA**- As we started studying life, it became apparent that life was not the same thing as DNA. Dead DNA and live DNA are similar in structure and forces on one verses the other would cause similar reaction.
- **Life was a different dimensional dynamo** from matter and forces on that matter. While it is different, nothing can be established in our universe unless life is instituted and a general acceptance by the combination of the conscious group must be assured before a true world can be generated. As people try to veer away from the common knowledge or viewpoint our world changes. When God stated that *"faith of a grain of mustard-seed would move mountains"* tells us how important our conscious minds are to our reality.
- **The Atrophic Universe Theory**- Tells us that each of us hold the world together. We cannot die or a piece of the universe will be lacking. Conservation of Energy won't allow it.
- **Life Between Lives**- The old Purgatory looks a little different than described in many Church Documents. Instead, substantial research has shown that people must learn more than they typically learn about the non-carnal existence so they reenter life several times.
- **God Exists**- Unfortunately, the universe cannot exist alone. A controller keeps everything from slipping back into entropy and disappearing. This God is the creator of everything and from the other books in the series; I hope that you understood how very important it is for the God stability to work in this

universe. In-waves and out-waves stabilize the non-living. God stabilizes the living/conscious elements of the universe.
- **God is Light**- In the form of the Holy Spirit or Holy Ghost; we found confirmation that this type of light was extremely important in not going to what we call Hell.
- **Death Isn't Death**-We finally looked at this thing we call death. We, possibly, can use death as an opportunity. There seems to be some? Chances to gain the "light" from the Holy Ghost that allows us to be transferred to another universe [Heaven], but no one knows how many chances and it seems that it is getting harder to become en-"Light"-ened by the Holy Ghost. If one gets the right credentials, one can be transferred to a better place.
- **Hell is Worse Than Bad**- We found that this whole lake of fire thing is not nearly as bad as not being part of existence that seems to be a truer definition of Hell.
- **Resurrection-** People who finally understand the "light" don't have to experience the nightmare of no stable universe and no interaction between entities as we can imagine this Hell to be.

With that, I must end this book.

About The Author

Steve Preston is a long time author of scientific, esoteric facts. His series on the creation of mankind is shown below. The series focuses on the painful truths rather than whitewashed details that make us comfortable. If you are interested in the truth instead of comfort, please continue to read and, while you are at it, review other works by Mr. Preston as shown below.

Four Part Series "Vibrational Matter"
- *Vibrational Matter*
- *10-Dimentional Universe*
- *Walk Though a Wall and Time*
- *The Meaning of Light and Life*

Eight Part Series "History of Mankind"
- *The First Creation of Man*
- *The Second Creation of Man*
- *The Creation Of Adam And Eve*
- *The Antediluvian War Years*
- *Man After the Flood*
- *Life After the Babel War*
- *A New View Of Modern History*
- *The 20th Century To The End Of Time*

Truth Series
- *The Truth About Dinosaurs*
- *The Truth About The Earth*
- *The 7 Destructions of the Earth*
- *The Truth About the Heaven War*
- *Truth About Dinosaurs*
- *Who Really Discovered the Americas?*
- *God Didn't Make The Ape*

Planet Series
- *When Did People Live on The Moon?*
- *Evolution of the Planets*
- *The Day Venus Exploded*
- *Living on Mars*

Odd Series
- *The Book Of Odd*
- *More Oddness*
- *Why Are There So Many Anomalies?*
- *Stupid Science*

Other Works
- *A Closer Look At Genesis*
- *A Closer Look At Lincoln*
- *Adam, Lilith, and Eve*
- *America's Civil War Lie*
- *Ancient History of Flying*
- *Behind the Tower of Babel*
- *The Funny Book of Law*
- *When Giants Ruled the Earth*
- *Lizard People*

www.ingramcontent.com/pod-product-compliance
Lightning Source LLC
Chambersburg PA
CBHW051646170526
45167CB00001B/353